THE GALACTIC CENTER
PROCEEDINGS OF THE SYMPOSIUM HONORING C. H. TOWNES

AIP CONFERENCE PROCEEDINGS 155

RITA G. LERNER
SERIES EDITOR

THE GALACTIC CENTER

PROCEEDINGS OF THE SYMPOSIUM
HONORING C. H. TOWNES
BERKELEY, CA 1986

EDITOR:
DONALD C. BACKER
UNIVERSITY OF CALIFORNIA

AMERICAN INSTITUTE OF PHYSICS   NEW YORK 1987

Authorization to photocopy items for internal or personal use, beyond the free copying permitted under the 1978 US Copyright Law (see statement below), is granted by the American Institute of Physics for users registered with the Copyright Clearance Center (CCC) Transactional Reporting Service, provided that the base fee of $3.00 per copy is paid directly to CCC, 27 Congress St., Salem, MA 01970. For those organizations that have been granted a photocopy license by CCC, a separate system of payment has been arranged. The fee code for users of the Transactional Reporting Service is: 0094-243X/87 $3.00.

Copyright 1987 American Institute of Physics

Individual readers of this volume and non-profit libraries, acting for them, are permitted to make fair use of the material in it, such as copying an article for use in teaching or research. Permission is granted to quote from this volume in scientific work with the customary acknowledgment of the source. To reprint a figure, table or other excerpt requires the consent of one of the original authors and notification to AIP. Republication or systematic or multiple reproduction of any material in this volume is permitted only under license from AIP. Address inquiries to Series Editor, AIP Conference Proceedings, AIP, 335 E. 45th St., New York, NY 10017.

L.C. Catalog Card No. 86-73186
ISBN 0-88318-355-2
DOE CONF-8610211

Printed in the United States of America

Contents

Preface ix

Invited Talks

The Stellar Cluster 1
    D. A. Allen
Dust Emission and the Evidence for Star Formation 8
    I. Gatley
Atomic and Molecular Gas in the Circumnuclear Disk 19
    R. Güsten
The Compact Nonthermal Radio Source 30
    K. Y. Lo
The Ionized Gas 39
    T. R. Geballe
Galactic Positron Annihilation Radiation 51
    R. Ramaty and R. E. Lingenfelter
The Evidence for and Against the Existence of Supermassive Black Holes in E Galaxies 62
    W. Sargent
The Central Object: Some Comments and Speculations 71
    M. J. Rees

The Stellar Cluster

The Stellar Population 79
    M. J. Lebofsky and G. H. Rieke
Kinematics of Individual Stars 83
    K. Sellgren, D. N. B. Hall, S. G. Kleinmann, and N. Z. Scoville
Stellar Kinematics in the Central 10 pc 87
    M. T. McGinn, K. Sellgren, E. E. Becklin, D. N. B. Hall, and I. Gatley
Is There a Cusp in the Stellar Distribution? 91
    G. H. Rieke and M. J. Lebofsky

The Circumnuclear Ring

The 18-cm OH Distribution 95
    A. Sandqvist, R. Karlsson, J. B. Whiteoak, and F. F. Gardner
NH_3 in the Molecular Ring 99
    J. M. Jackson, P. T. P. Ho, and A. H. Barrett
Hat Creek Aperture Synthesis Observations 103
    R. Güsten, R. Genzel, M. C. H. Wright, D. T. Jaffe, J. Stutzki, and A. Harris
Rotating Molecular Ring 106
    N. Kaifu, M. Hayashi, J. Inatani, and I. Gatley
The CO (J=2−1) Distribution in the Inner 10 pc 110
    Y. Fukui and E. Churchwell
CS Emission from the Galactic Center Ring 114
    N. J. Evans

Excitation Gradient of the Molecular Gas 118
    J. B. Lugten, G. J. Stacey, A. I. Harris, R. Genzel, and C. H. Townes
Mapping of C^+ Far-Infrared Emission 123
    J. B. Lugten, R. Genzel, M. K. Crawford, and C. H. Townes

THERMAL AND NONTHERMAL EMISSION

Radio Emission from Sgr A and Its Extended Halo 127
    M. Morris and F. Yusef-Zadeh
86-GHz Aperture Synthesis Observations 133
    M. C. H. Wright, R. Genzel, R. Güsten, and D. T. Jaffe
Small-Scale Structure in the Far-IR 138
    D. F. Lester, M. Joy, P. M. Harvey, and H. B. Ellis, Jr.
Observations of Galactic Center Gas Dynamics with a Cryogenic Echelle
Spectrometer 142
    J. H. Lacy, D. F. Lester, J. F. Arens, M. C. Peck, and S. Gaalema
8.3 and 12.4 Micron Imaging 146
    D. Y. Gezari, R. Tresch-Fienberg, G. G. Fazio, W. F. Hoffmann, I. Gatley,
    G. Lamb, P. Shu, and C. McCreight
Brackett Alpha Images 153
    W. J. Forrest, M. A. Shure, J. L. Pipher, and C. E. Woodward
Preliminary Results of the Spacelab-2 Infrared Telescope Survey of the
Galactic Plane at 2.4 µm 157
    G. J. Melnick, G. G. Fazio, D. G. Koch, G. H. Rieke, E. T. Young,
    F. J. Low, W. F. Hoffmann, and T. N. Gautier

ASTROMETRY

Position of IRS-7 162
    E. Becklin, H. Dinerstein, I. Gatley, M. W. Werner, and B. Jones
Proper Motion of the Compact, Nonthermal Radio Source 163
    D. C. Backer and R. A. Sramek
The Distance to the Center of the Galaxy 166
    J. M. Moran, M. J. Reid, M. H. Schneps, C. R. Gwinn, R. Genzel,
    D. Downes, and B. Rönnäng
The Next Series of Lunar Occultations, 1986–1989 168
    A. Sandqvist

RELATIONSHIP TO EXTERNAL GALAXIES

Low-Luminosity Seyfert Nuclei in Nearby Galaxies 172
    A. V. Filippenko and W. L. W. Sargent

ASTROPHYSICAL MODELS

A Magnetic Loop Model for Activity 176
    J. Heyvaerts, R. E. Pudritz, and C. A. Norman
Further "Loss of Weight" by a Black Hole 181
    L. M. Ozernoy
Positron Annihilation in the Galactic Center: "Chesire Cat" Compton
Scattering and "Excess Continuum" 184
    M. L. Bildsten and W. H. Zurek

GALACTIC CENTER ARC AND LOBES

Molecular Gas Associated with the Galactic Center Arc 188
    E. Serabyn and R. Güsten
Spatial and Kinematic Structure of the Thermal Components of the Galactic Center Arc 190
    F. Yusef-Zadeh, M. Morris, and J. H. van Gorkom
Evidence for Activity at the Galactic Center Based on Low-Frequency, Radio-Continuum Observations 196
    N. E. Kassim, W. C. Erickson, and T. N. LaRosa

Author Index 203

Preface

On 25 October 1986 a one-day Symposium on the Galactic Center was held in the Bechtel Auditorium on the campus of the University of California at Berkeley. The Symposium was organized to honor the recent (formal) retirement of Professor Charles H. Townes from the Physics Department. Reinhard Genzel, Jack Welch, and I were co-organizers of the Symposium. More than 150 scientists from around the globe attended the meeting. The day's invited talks and poster contributions focused our attention on activity in the central 10 parsecs of the Milky Way galaxy. This topic was chosen since Professor Townes and his co-workers at Berkeley have been continually engaged in revealing the secrets of the very central region of our galaxy as they developed new infrared techniques for astrophysical investigations.

This volume begins with a record of the invited talks. A panel—C. Townes, G. Rieke, E. Becklin, and P. Mezger—led a discussion at the end of the day that reviewed the posters and explored issues that arose during the day. There is no record of this panel discussion. The 36 posters that were on display in the lobby surrounding the auditorium were more than even an expert could digest in a single day. This volume provides us all with excellent summaries of the new data and discussions in these posters. The poster summaries follow the invited talks, and are organized by subject matter.

The most confusing jargon in this Proceedings concerns material surrounding the peak of the stellar distribution; it is called ring cloud, molecular disk, neutral ring, nuclear torus, rotating torus, circumnuclear ring, and possibly other names. This list describes what is found at a radius of about 2 pc. Güsten's review introduces this topic and there are seven poster summaries on observational aspects. Another area of confusion has to do with larger structures: lobes, filaments, and arcs. These are discussed in a few posters, but were outside the main focus of the day. Since there are many excellent programs investigating these structures in parallel, some nomenclature standards need to be developed before the next galactic center meeting, or else we will spend a lot of time talking in circles.

The conference participants began their discussions at a reception Friday evening in the Physics Department Lounge in LeConte Hall. Saturday evening many of the participants joined the Physics Department in a retirement banquet at the Faculty Club.

The Townes Symposium would not have been possible without the combined support from the Physics Department, the Radio Astronomy Laboratory, the College of Letters and Sciences, and the Space Sciences Laboratory. People's time and effort count even more than hard cash. The Symposium owes its thanks to Jaqueline Mundo and Fern Shugart from the Physics Department and Sasha Paulsen and Pat Crowder from the Astronomy Department and Radio Astronomy Laboratory. This Proceedings would not have been possible without the constant efforts of Sasha Paulsen.

Donald C. Backer

THE STELLAR CLUSTER

D.A. Allen
Anglo-Australian Observatory, PO Box 296, Epping 2121,
N.S.W., Australia

ABSTRACT

I review our present knowledge of the stellar distribution within the inner parsec of the Galaxy.

INTRODUCTION

Any mature galaxy, such as we inhabit, exhibits a pronounced peak of brightness towards its dynamical centre. There its predominantly population II stars congregate most densely, roughly according to the prescription of a King model[1]. We therefore expect to find within our Galaxy a cusp-like concentration of unresolved stars, peaking near the radio and far-infrared centres of activity in the constellation Sagittarius.

Because we lie so close to the symmetry plane of our Galaxy, that population II cusp must be observed through more than 7 kpc of interstellar gas and dust, a Galactic smog which causes a total visible extinction of some 30 magnitudes. Although the density of stars is extreme within the central parsec, it is not so extreme as to overwhelm that attenuation factor of 10^{12}. To the uninitiated observer relying on visible radiation, the Galaxy has no centre; were this not so we should undoubtedly find references in reviews such as this dating back to the late 17th century, when that gifted young English astronomer Edmond Halley made the first survey of the southern sky from the island of St. Helena. As it is, the earliest reference to the population II stellar concentration is dated 1968.

THE POPULATION II CUSP

By observing at a wavelength of 2.2 µm (the K window), Becklin and Neugebauer[2] mapped the central regions of the Galaxy at coarse spatial resolution, and located the stellar cusp quite unambiguously. At K the interstellar extinction is less than 3 magnitudes so that both the diffuse cusp and a number of discrete sources are totally dominant over foreground objects.

Interest turned to longer infrared wavelengths; remarkably, seven years elapsed before the central regions were again studied at K, with higher spatial resolution and by the same authors[3]. The new maps isolated many more discrete sources superimposed on the population II cusp, and a score of these sources have now been given designations. Of these the two most important are IRS 7, a particularly luminous M supergiant, and IRS 16, of which more anon.

Bailey[4] integrated the contours presented in the two Becklin and Neugebauer papers and showed that the underlying stellar profile showed a dependence of the surface brightness S(r) on radius r of the form:

$$S(r) \propto r^{-0.8} \qquad (1)$$

over nearly three orders of magnitude in r. The corresponding volume density N(r) is therefore:

$$N(r) \propto r^{-1.8} \qquad (2)$$

and the index is close to the values expected for a fully relaxed isothermal sphere (-2) or for the cusp around a very massive single object (-1.75 if relaxed, -1.67 if adiabatic).

No such cusp can continue to the singularity at r=0; an inner bound r_{core} can be defined, as in King models, and in a simplistic way we can consider S(r) constant for r < r_{core}. Bailey sought to show that r_{core} lay within the regime covered by the observations. However, the only datum he could find that gave any evidence for a flattening of the slope (see Fig. 1) lay wholly within the 2.5 arcsec beam used by Becklin and Neugebauer[3], and was therefore a measure of their beam profile rather than of the stellar cluster.

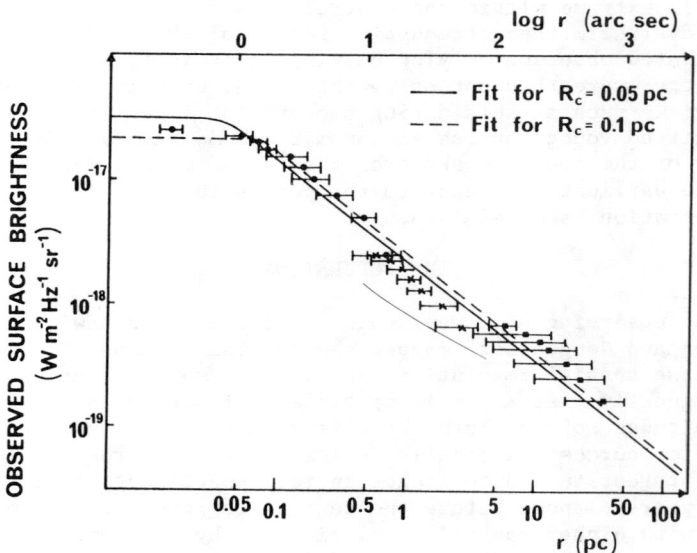

Fig. 1. The surface brightness S(r) over the inner few parsecs as derived by Bailey[4]. The innermost point, vital to the model fits, should be discarded.

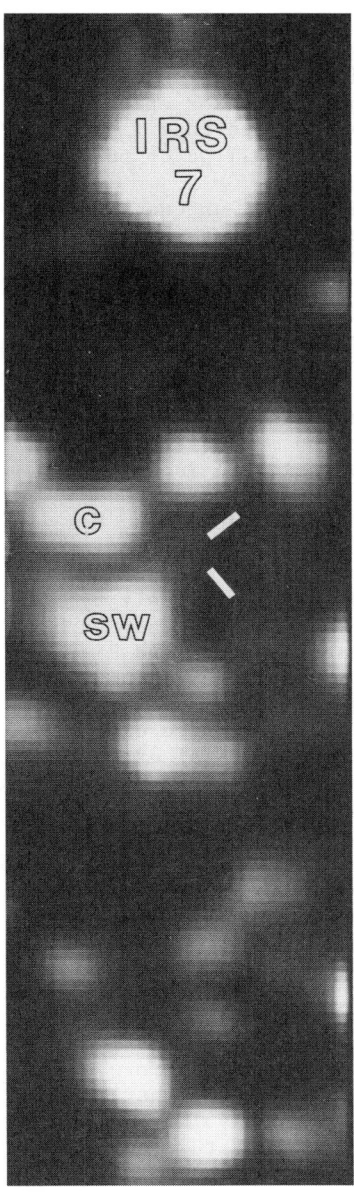

Allen, Hyland and Jones[5] pressed to still higher spatial resolution, and again resolved more point sources superimposed on the cusp. At that time the accumulated wisdom was that:

. IRS 16 was the true centre of the Galaxy and coincided with the compact non-thermal radio source Sgr A* discovered by Balick and Brown[6];

. the discrete sources comprised M supergiants, M giants and some dust-rich, compact H II regions.

Working on these premises Allen et al. removed the contributions of the more prominent point sources from their data and found no hint that r_{core} exceeded 0.5 arcsec (0.02 pc). Taken at face value these data suggested that a single point mass within IRS 16 dominated the central dynamics of the Galaxy.

Data gathered over the last few seasons have encouraged a different view of the Galaxy's core. Already in 1983 there was a suspicion that Sgr A* did not coincide with IRS 16, which itself comprised at least three components of unusually blue colour[7]. More recent data confirm the displacement: Allen and Sanders[8] find that Sgr A* coincides with a discrete source to the west of IRS 16 Centre whose apparent K magnitude is 11.4 in a beam of diameter 0.7 arcsec. Their high-resolution K image is reproduced in Fig. 2.

Fig. 2. The highest spatial resolution yet obtained in the Galactic centre[8] shows some of the components of IRS 16 (SW, C) to be extended, and a probably discrete source (arrowed) coincident with Sgr A*. The image measures 18x5 arcsec.

Sgr A* remains the one candidate for a black hole at the Galactic centre, and indeed is difficult to explain in any other way, as Lo demonstrates later in this volume. Moreover, the limits on the proper motion of Sgr A* found by Backer et al. (poster paper presented at this Workshop) indicate that if a black hole of mass $\sim 10^6$ $M_\odot$ exists within the central 0.1 pc of our Galaxy then it must be identified with Sgr A*.

IRS 16 itself is also highly unusual, if not unique. Its components are all much bluer than any other sources in the region. Associated with them is He I 2.058 μm emission[9] normally found only in the vicinity of the exciting stars of H II regions. Moreover, Doppler broadening by ~1000 km/sec is found in both the He I[10] and Brackett lines[9] in the vicinity of IRS 16. It does not seem possible to identify any significant part of IRS 16 with population II stars.

It is thus now clear that our understanding of the Galactic centre is limited by confusion. There are so many discrete objects that it is difficult, if not impossible, to isolate the underlying core of population II stars. Integration of the observed brightness maps within narrow annuli to determine $S(r)$ is not appropriate because at small r the discrete objects dominate. An alternative approach is to use the lowest brightness seen within each annulus, on the basis that this is most likely to sample the unresolved K-M giants many of which must occupy each pixel of present data. That scheme was used by Rieke and Rieke (poster paper presented at this workshop), who inferred the surprisingly large value r_{core} ~ 20 arcsec.

Rieke and Rieke's method will fail in the presence of patchy extinction. They themselves identify small patches of extinction; their conclusions will be in error if the extinction is unusually high in the dark area to the west of Sgr A* (see Fig. 2). For the present, 20 arcsec should probably be regarded as an upper bound to r_{core}.

A theoretical lower bound to r_{core} in the cusp around a massive object is about 1 arcsec. Within that radius, star-star collisions destroy all but a small number of stars (Sanders, private communication). Within the 0.7 arcsec beam used in ref. 7 the peak surface brightness for r_{core} = 1 arcsec would be K=11.3, which corresponds to the source coincident with Sgr A*. On Fig. 2 that source is smaller than 2 arcsec diameter, and is considerably brighter than the region to its west; it does not therefore appear possible to identify it with the population II cusp either. This is an argument that r_{core} > 1 arcsec, but it is a valid argument only if Sgr A* is the dynamical centre. A core of this surface brightness could underlie IRS 16. It must be concluded that there is no good observational lower bound to r_{core}, and the only safe conclusion at present is that:

$$1 < r_{core} \leq 20 \text{ arcsec}$$
$$\text{or} \quad 0.04 < r_{core} \leq 0.8 \text{ parsec}$$

for the IAU's currently adopted distance of 8.5 kpc to the Galactic centre.

The prospect of improving this unsatisfactory situation is quite good. Imaging spectrometers now exist which can examine the CO bandheads in the 2.3 μm region throughout, say, the inner 40 arcsec. Hot stars such as IRS 16 will show no CO absorption, while discrete M giants and M supergiants will generally have more intense CO features than the unresolved somewhat earlier type giants. Thus we can potentially isolate by means of broad band and spectroscopic imaging the different stellar types and the patchwork effects of extinction. Fig. 3 (a and b) illustrates a first step towards this goal, from AAT data taken in 1986 by Allen, Bailey, Hillier, Hyland and Roche.

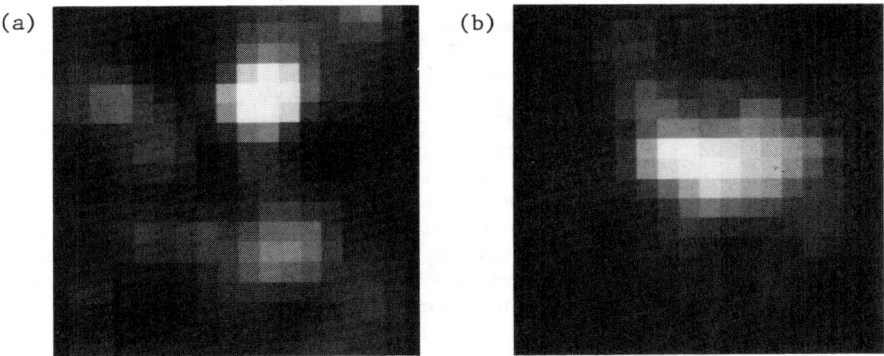

Fig. 3. Images at 3.5 arcsec resolution of the inner 30 arcsec of the Galaxy which use the CO absorption bands to separate sources (a) with and (b) without the CO absorption bands expected of M supergiants and population II stars.

THE POPULATION I CUSP

If Rieke and Rieke are correct that r_{core} for the population II stars is 20 arcsec, then we must accept that superimposed on that distribution is a second population of hot stars and M supergiants with a much smaller core radius. These are the stars which dominated in Bailey's analysis. Their distribution can be regarded as a population I cusp, though since we are now dealing with a much smaller number of more luminous stars the concept of a smooth cusp may not be appropriate. That the most recently formed stars occupy a more centrally condensed distribution is not necessarily surprising provided that there has not been time for them to relax in the potential of the population II stars.

The population I cusp clearly focuses our attention on IRS 16, which lies near its geometrical centre.

IRS 16

As noted above, IRS 16 is an unusual object. It is most readily explained as a cluster of hot stars inhabiting and ionizing an H II region which is, for some reason, dust free. The associated broad emission lines need be nothing more exotic than mass loss from OB or Wolf-Rayet stars, and it is tempting to invoke the same stars as the source of most of the Lyman continuum photons, and indeed of the luminosity of the Galactic core.

At present there is conflicting evidence on the morphology of the individual components of IRS 16. The conflict is between two different types of observation both of which are technically very difficult. Mapping with a 0.7 arcsec aperture[8] (Fig. 2) shows three of the four components - IRS 16 NE, Centre and SW - to be extended. On the other hand, observations of a lunar occultation of this region, presented by Adams et al. as a poster paper at this Workshop, suggest that only IRS 16 NE is extended. In the case of the occultation measurement, the signal/noise ratio is barely enough to show that all the radiation from each component is accounted for by a discrete disappearance of the signal. The images, on the other hand, were built up by a raster scan using a single detector, and could have been influenced by seeing.

If the components are indeed individual stars, and with the knowledge that they do not exhibit CO absorption, then they must lie above the main sequence and be hotter than G stars. The most luminous Wolf-Rayet stars, the Galactic hypergiants (such as HD 316285), η Carinae, and the B supergiants are the only known candidates. All are rare objects, and a cluster of them at the Galactic centre must be regarded as surprising.

A less surprising interpretation, thus favoured by Occam's razor, is that the components of IRS 16 are all compact clusters of hot stars. We require that their integrated K magnitude be 8.0, the total flux of all the blue components of IRS 16. If they comprise exclusively stars of one spectral subclass, clearly an untenable simplification, then we should derive the following properties:

Spectral type	O5	O8	B0
Number of stars	55	125	250
Luminosity ($L_\odot$)	10^7	10^7	8×10^6
Lyman continuum photons/sec	4×10^{50}	10^{50}	10^{49}

Each component of IRS 16 would contain 20-30% of the number of stars given here. For comparison, the total luminosity inferred from the far-infrared dust emission, and using an estimate for the solid angle subtended by that dust, is $(1-3) \times 10^7$ $L_\odot$; the Lyman flux required to ionize the region is estimated at 2×10^{50} photons/sec[12].

We see a deliciously consistent solution if we make all the stars mid O. Unfortunately, we appear to be prohibited from using this solution because neither [S IV] nor [Ar III] emission is seen in the region, implying a radiation field more representative of late O or early B stars. Since these figures are rather loose, and since IRS 16 may not solely account for the luminosity or the ultraviolet heating, a solution along these lines may be possible.

I take comfort in the fact that those who favour a single object as the font of most of the energy at the Galactic centre find even greater difficulties working within these constraints.

It remains to be noted that if the components of IRS 16 are stellar clusters, then their very existence places a severe upper limit on the mass of any black hole at the location of Sgr A*, or indeed elsewhere within a few arcsec, of about $10^3 M_\odot$, as argued by Allen and Sanders[8]. Then, using arguments about the growth habits of black holes[13,14] we can further reduce the inferred mass to $10^2 M_\odot$. The same argument requires that IRS 16 resides in an essentially flat potential field of the population II cusp, and hence that r_{core} be large enough to contain most of IRS 16. This argument possibly raises the lower bound or r_{core} from 1 to 1.5 arcsec.

It is a pleasure to acknowledge the stimulating insights provided by Roger Blandford.

REFERENCES

1. I. R. King, Astron. J. 71, 64 (1966).
2. E. E. Becklin and G. Neugebauer, Astrophys. J. 151, 145 (1968).
3. E. E. Becklin and G. Neugebauer, Astrophys. J. 200, L71 (1975).
4. M. E. Bailey, M.N.R.A.S. 190, 217 (1980).
5. D. A. Allen, A. R. Hyland and T. J. Jones, M.N.R.A.S. 204, 1145 (1983).
6. B. Balick and R. L. Brown, Astrophys. J. 194, 265 (1974).
7. J. W. V. Storey and D. A. Allen, M.N.R.A.S. 204, 1153 (1983).
8. D. A. Allen and R. H. Sanders, Nature 319, 191 (1986).
9. T. R. Geballe, K. Krisciunas, T. J. Lee, I. Gatley, R. Wade, W. D. Duncan, R. Garden and E.E. Becklin, Astrophys. J. 248, 118 (1984).
10. D. N. B. Hall, S. G. Kleinmann and N. Z. Scoville, Astrophys. J. 260, L53 (1982).
11. E. E. Becklin, I. Gatley and M. W. Werner, Astrophys. J. 258, 135 (1982).
12. R. L. Brown and H. S. Liszt, Ann. Rev. Astron. Astrophys. 22, 223 (1984).
13. J. G. Hills, Nature 254, 298 (1975).
14. L. M. Ozernoy, I.A.U. Colloq. 45, 121 (1977).

DUST EMISSION AND THE EVIDENCE FOR STAR FORMATION

Ian Gatley
National Optical Astronomy Observatories, Tucson, Az. 85726*

ABSTRACT

Dust obscures our view of the Galactic center, and complicates enormously the search for a "central engine". One straightforward but indirect method is to study the thermal emission from dust in the nucleus. A very simplistic assumption, that the nucleus is choked with dust, leads to the prediction that a central engine, if present, will produce a single bright infrared source. Observations made more than a decade ago excluded this naive possibility. Instead, $10\mu m$ images of the Galactic center are complicated, with multiple peaks in the emission.

Existing radio observations of Sgr A had already suggested the presence of ultraviolet radiation in the nucleus, and so it was that some workers saw the $10\mu m$ image as evidence for a burst of star formation. The demonstration of the existence of late type supergiants within the field of the $10\mu m$ map encouraged that interpretation.

The possibility that the dust density in the inner Galaxy is actually very low was not taken seriously until it was directly demonstrated by far infrared observations. These observations showed that a ring of neutral material encircles the nucleus at a radius of 2 parsecs, that within the central cavity of this ring the dust density is low, and that this inner region is transparent to optical and ultraviolet radiation; the structure observed at $10\mu m$ is located within the cavity in the ring.

Maps of the infrared color temperature distribution are symmetric and peak centrally in the vicinity of the nuclear source IRS16. There are no temperature peaks at the $10\mu m$ brightness peaks. The energetics of the inner few parsecs of the Galaxy are dominated by a strong source of luminosity resident at the Galactic center.

Wisps and streamers of material falling inward are exposed to the radiation field of the central object, which ionizes the gas and heats the dust. The clumpy density distribution is responsible for the complicated appearance of the $10\mu m$ map.

There is no direct or compelling evidence for star formation in the very center of the Galaxy. The appearance of the interstellar medium in the inner Galaxy is dictated by the presence of a nuclear source. We cannot presently tell if the nuclear source is powered by star formation or by some more exotic object, such as a black hole.

*Operated by the Association of Universities for Research in Astronomy, Inc. under contract with the National Science Foundation.

INTRODUCTION

We study the center of our Galaxy to obtain detailed information about conditions in the nucleus, but practical and technical difficulties limit and distort our understanding. We are constrained chiefly to radio and infrared observations by the large amount of dust in the line of sight to the Galactic center. Technical developments in these observational disciplines change or refine our picture of the nucleus, and even the order in which these developments are made can bias our interpretation.

Early radio observations identified the HII region Sgr A, and so suggested the presence of a source of ultraviolet luminosity in the Galactic center[1,2]. A simple method by which to study a luminous source in a dusty environment is to measure the thermal emission from the dust. In this way the total luminosity and location of the sources which heat the dust may be determined.

The energy distribution of the thermal emission from dust in the Galactic center peaks in the far infrared, a wavelength region which is only observable from airborne platforms. Figure 1 show an early far infrared map[3] made from a balloon-borne telescope, in which the peaks are labelled by the names of the corresponding radio sources.

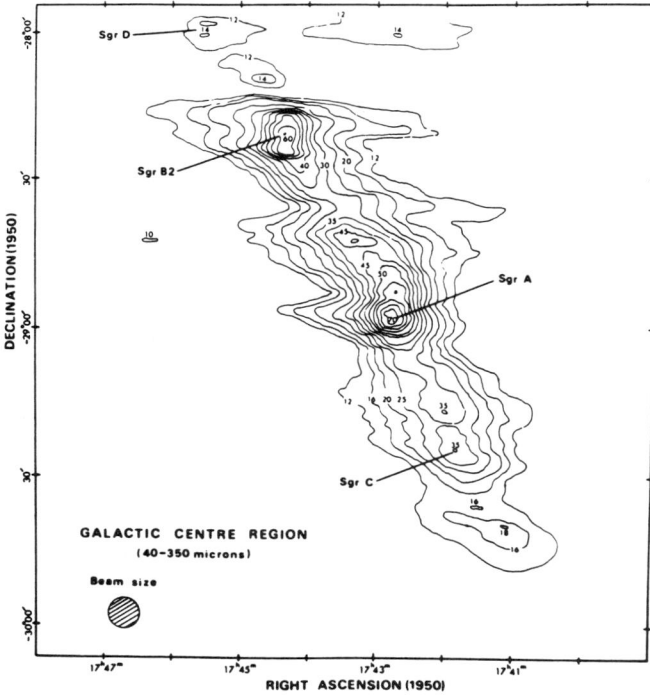

Fig. 1 A far infrared map of the Galactic center region.

In the mid 1970's neither radio nor far infrared observations could be made with sufficient angular resolution to study the detailed structure of Sgr A. It was, however, well known that 10μm images of HII regions often bore a striking resemblance to radio images[4]. Furthermore, 10μm studies of HII region complexes in the spiral arms of the Galaxy often revealed bright, compact sources identified as sites of recent star formation. These sources were typically deep within their parent molecular cloud. Thus the infrared community was widely familiar at this time with a situation in which 10μm maps of HII regions revealed the locations of luminosity sources deeply embedded in neutral material.

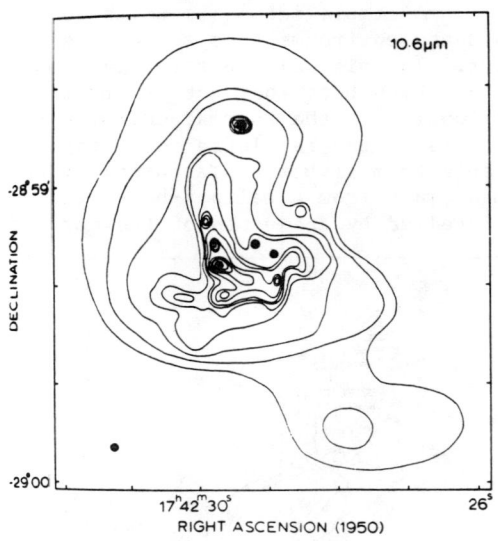

Fig. 2. A 10μm map of the Galactic center[5].

This perspective, together with a common prejudice that the inner Galaxy <u>must</u> be very dusty, made it inevitable that the complex structure in the 10μm image of the Galactic center would be interpreted as evidence for multiple luminosity sources. The very possibility of a central engine now seemed remote.

It became the fashion to view the Galactic center as unremarkable[5]; Sgr A was just another HII region, and the nucleus was a disordered aggregation of familiar phenomena. The HII region was powered by recently formed stars. The case for star formation gained support with the identification of late-type supergiants in the field of the 10μm map[6].

The first far infrared observations with enough angular resolution to determine structural details in Sgr A were made from the Kuiper Airborne Observatory[5,7]. A startling and unexpected discovery was made; the dust density in the central parsec of the Galaxy is very low[8]. This discovery restored the credibility of models with a single source of luminosity[9], and renewed the debate between central engine and star formation.

HEATING THE DUST

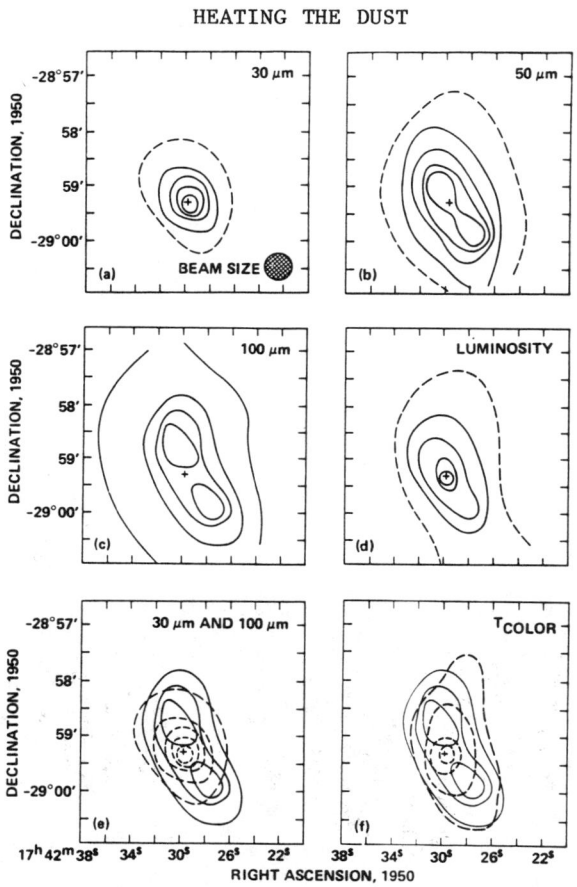

Fig. 3. Far infrared maps of the Galactic center.

The data of figure 3 have been discussed in detail previously[8,9,10]. Quoting from an earlier work[10], "figure 3a, b, c shows maps of the Galactic center with 30" resolution at wavelengths of 30μm, 50μm and 100μm. The position of the Galactic center source IRS16 is indicated by a cross in each map. At 30μm the emission is compact and centered on the Galactic nucleus, but at longer infrared wavelengths the emission takes on a double-lobed appearance, with the lobes extended along the plane of the Galaxy.

The peaks in color temperature and luminosity at the position of the Galactic center suggest that the dust is heated primarily by a centrally concentrated source of luminosity, but the central depression in 100μm surface brightness implies that there is a local minimum in the dust density in the central parsec of the Galaxy.

The double-lobed appearance of the emission at longer wavelengths arises naturally if the dust density is highest in the Galactic plane. Non-ionizing radiation streaming out of the central parsec (where the dust density is evidently too low to absorb it) enters a region of higher dust density, which is shaped like a ring lying in the Galactic plane. Seen in projection, the emission from this ring causes the double lobes".

The structure seen in the 10μm map (figure 2) is all located within the central cavity in this ring. The transparency of this region to optical and ultraviolet radiation allows the possibility that a single source of luminosity powers the whole complex.

Now, the interpretation of the far infrared data relies simply on the fact that peaks in maps of thermal emission from dust are either caused by a local increase in dust temperature or by local increase in dust density. The issue of the origin of the peaks in the 10μm map can be attacked[10] by the consideration of this same idea.

In an earlier review[10], it was demonstrated from consideration of the color temperature distribution that the appearance of the 10μm map is primarily caused by density variations. Figure 4 shows some results from that work not previously included in a written presentation[11].

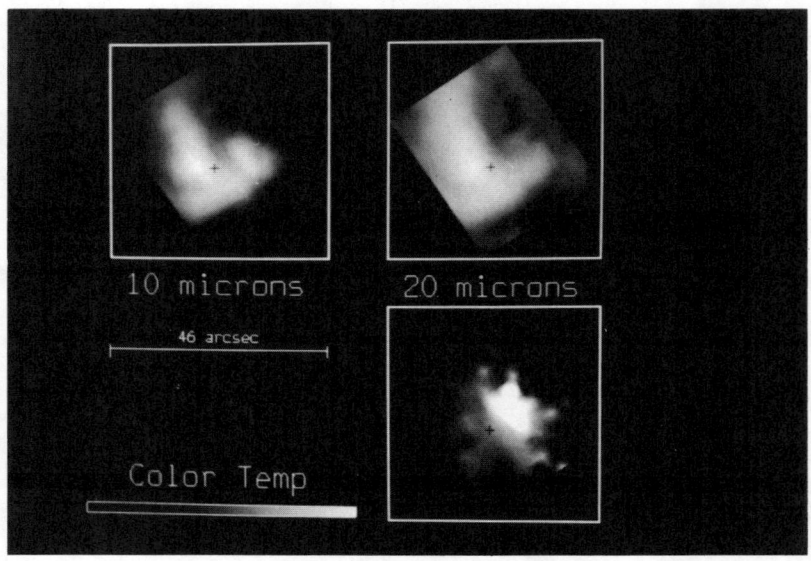

Fig. 4a

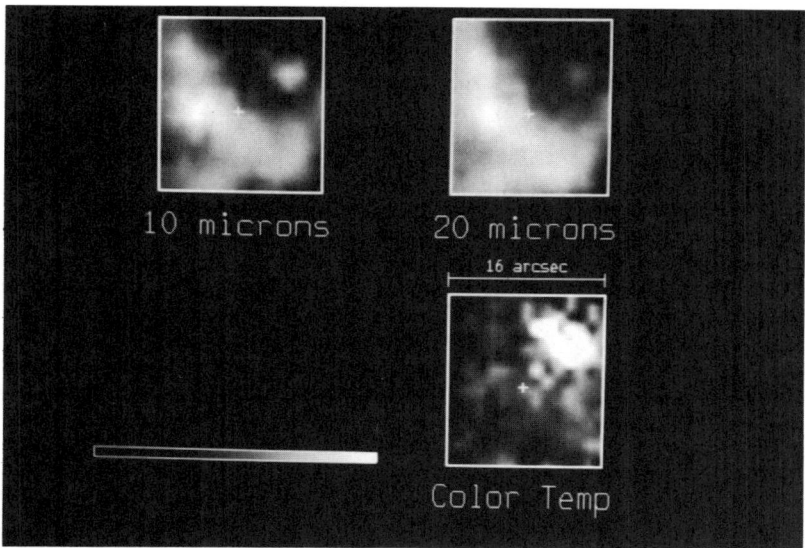

Fig. 4. Surface brightness and color temperature distribution in the Galactic center.

The grey scale intensities in figure 4 are proportional to the logarithm of the surface brightness in the 10μm and 20μm images and to the ratio of these surface brightnesses in the color temperature images. The spatial resolution is 4" in figure 4a and 2" in figure 4b.

The failure of the color temperature to peak on the bright positions in the infrared "ridge" is a striking refutation of the notion that the 10μm peaks are primarily self-luminous. The stellar sources IRS3 and IRS7, located northwest of IRS16 (shown by a cross), do appear as hot spots in the color temperature images. After subtraction of these hot spots the color temperature distribution is approximately circularly symmetric about IRS16.

Sadly, these results met with a hostile reception[12] at their first presentation, and their veracity was cast into doubt by the exhibition of a sketch which has sometimes been mistaken for observational data[12].

Since that time another experiment has confirmed that none of the 10μm ridge sources is heated appreciably by internal luminosity sources. Figure 5 shows the color temperature distribution and figure 6 the dust opacity distribution derived from measurements with the Goddard infrared array camera[13].

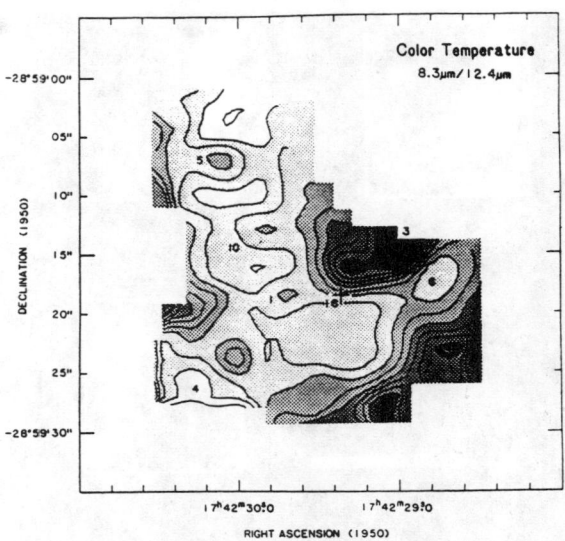

Fig. 5. Derived color temperature for the Galactic center source complex, calculated from the 8.3μm/12.4μm image ratio assuming blackbody spectra. Darker shades are higher temperatures. The lowest temperature contour is T_c=220K; contours are plotted at 20K intervals. The peak value T_c=400K occurs at IRS3. The value at IRS1, 5 and 10 is 290K, and at IRS6 260K.

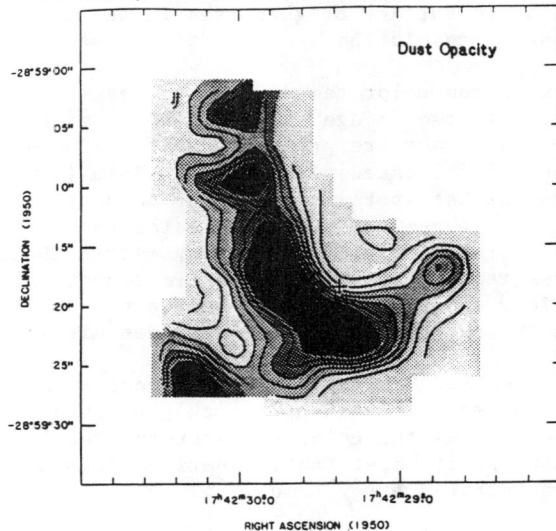

Fig. 6. Derived opacity of warm dust grains giving rise to the observed Galactic center 8.3μm emission. Darker shades are high opacity. The position of IRS16 (center) is shown by the cross. The lowest contour plotted is $\tau = 5.0 \times 10^{-4}$, and the contour interval is $\tau = 5.0 \times 10^{-4}$. The peak opacity, $\tau = 5.6 \times 10^{-3}$, occurs at IRS1.

These results are consistent with any model which proposes that a strong source of luminosity resident at the Galactic center dominates the energetics of the inner few parsecs of the Galaxy.

THE STRUCTURE OF SGR A

The existence of an HII region in the Galactic nucleus is a clear signpost of activity, and the vast majority of observational research yet performed in the study of the Galactic center is concerned with the appearance of Sgr A. We search for, and find, fascinating microstructure, the correct interpretation of which may be crucial in determination of physical conditions in the nucleus.

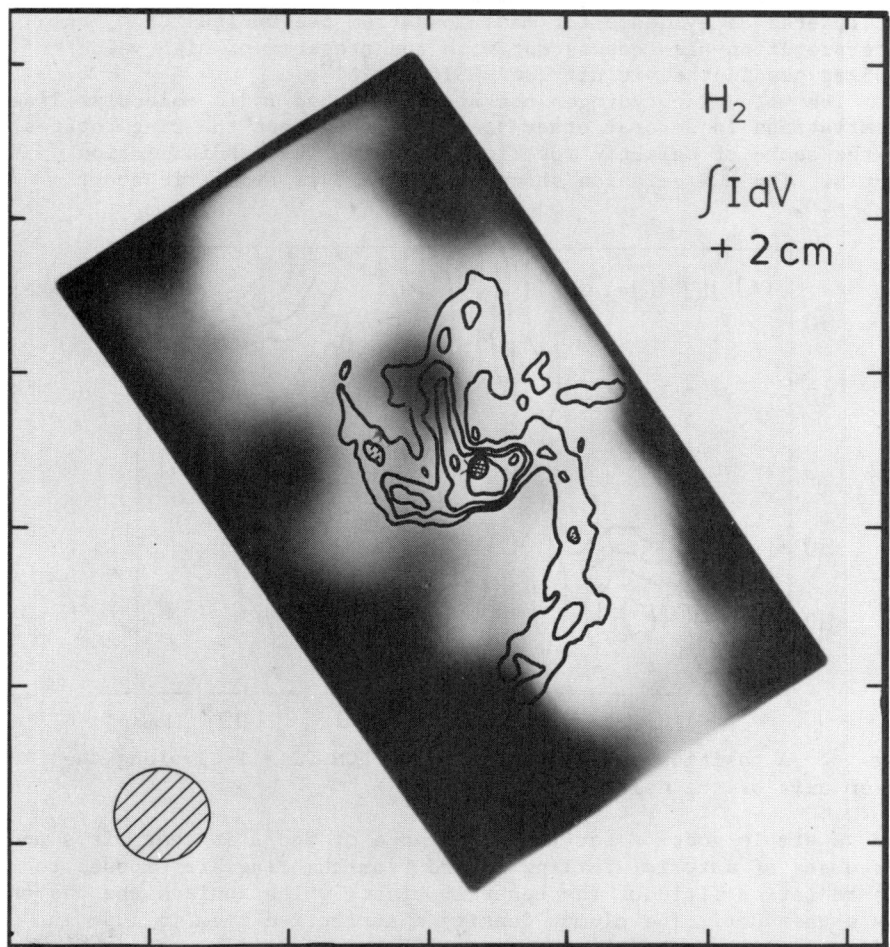

Fig. 7. Emission from vibrationally excited molecular hydrogen (grey scale) and ionized hydrogen (contours) in Sgr A.

The evidence provided by the thermal emission from the dust suggests that this microstructure may result simply from the interaction of a single central engine with the nearby, rather tenuous, interstellar medium. The demonstration of symmetries in the inner Galaxy is the logical opposite to the emphasis on disorder which motivated the early case for star formation.

The symmetric double-lobed appearance of the far infrared emission (figure 3) is interpreted as evidence for a ring of neutral material which lies in the plane of the Galaxy and encircles the nucleus. The possibility that the nuclear source creates the central cavity in this ring motivated a search for shock excited molecular hydrogen emission at the locus of the ring[14].

The molecular hydrogen emission shown in figure 7 has been interpreted as evidence for mass loss from the nucleus[14,15], an interpretation also consistent with the presence of high velocity ionized gas in the vicinity of IRS16[16,17,18].

The molecular hydrogen observations[15] and radio molecular line observations in several other species[19] show that the ring rotates in the sense of Galactic rotation, and also that radial motion occurs. The HCN emission shown in figure 8 is symmetric about $V_{LSR} = 0$.

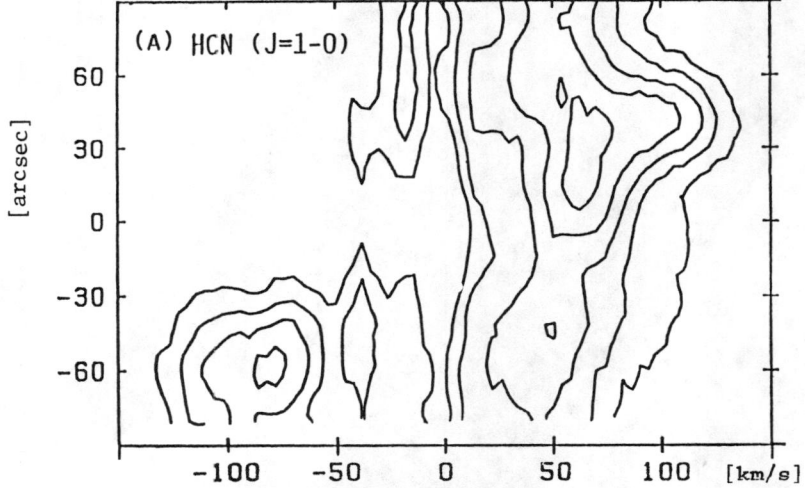

Fig. 8. A position-velocity diagram in HCN (J = 1-0) along the major axis of the Galaxy.

A simple model[15] for the appearance of Sgr A is that wisps and streamers of material falling inward from the ring are exposed to the radiation field of the central object, which ionizes the gas and heats the dust. The clumpy density distribution seem in 10μm and radio maps is a natural consequence of this accretion through the mass loss wind from the nucleus. The mass loss and accretion rates are comparable, and the mean density of the central parsec is low.

In this model the late-type supergiant IRS7, frequently cited as important evidence for star formation, cannot be within Sgr A. If it were, then its outer envelope would be ionized by radiation from the central source, and it would appear as a feature in the radio free-free map[20].

THE NATURE OF THE CENTRAL SOURCE

The observations of Sgr A are simply and consistently explained as the result of interactions between a nuclear source and the nearby interstellar medium. This nuclear source is located in the vicinity of IRS16, but it is not uniquely identified. Almost all of the observational evidence is consumed in arriving at this perspective. So it is that both "central engine" and "star formation" models are presently viable.

The earlier arguments advanced in favor of star formation are specious, and the arena in which to construct starburst models has shrunk almost to the limits of our observational capabilities. Yet IRS16 does show structure[21,22]. Further refinements in observational strategy and capability are sure to tighten the constraints on star formation models. The general challenge to such models is to provide a plausible scenario for the birth, evolution and death of stars consistent with the structure observed in Sgr A. Central engine models simply explain the structure of Sgr A, and can also accomodate the existence of the peculiar radio source[23] Sgr A* in a natural way.

ACKNOWLEDGEMENTS

I am extremely grateful to my collaborators Eric Becklin, Charles Telesco, Norio Kaifu and Masahiko Hayashi, to George Rieke and Nick Scoville, who sharpened my perspective on this problem, and to Ms. Linda Africano, who prepared the manuscript.

REFERENCES

1. D. Downes, A. Maxwell, Ap. J., 146, 653 (1966).
2. R. D. Ekers, W. M. Goss, U. J. Schwarz, D. Downes, D. H. Rogstad, Astr. Ap, 43, 159 (1975).
3. J. A. Alvarex, I. Furniss, R. E. Jennings, K. J. King, A. F. M. Moorwood, HII Regions and the Galactic Centre, Proceedings of Eighth ESLAB Symposium, ESTEC, Noordwijk, Netherlands, p. 69 (1974).
4. E. E. Becklin, C. G. Wynn-Williams, Publ. Astron. Soc. Pacific, 86, 5 (1974).
5. G. H. Rieke, C. M. Telesco, D. A. Harper, Ap. J., 220, 556 (1978).
6. G. Neugebauer, E. E. Becklin, S. Beckwith, K. Matthews, C. G. Wynn-Williams, Ap. J. (Letters), 205, L139 (1976).
7. P. M. Harvey, M. F. Campbell, W. F. Hoffmann, Ap. J. (Letters), 205, L69, (1976).

8. E E. Becklin, I. Gatley, M. W. Werner, Ap. J., <u>258</u>, 135 (1982).
9. I. Gatley, E. E. Becklin, IAU Symp. No. 96, p. <u>281</u> (1981 .
10. I. Gatley, AIP Conf. Proc. No. 83, p. 25 (1982).
11. I. Gatley, E. E. Becklin, C. M. Telesco, unpublished.
12. G. H. Rieke, M. J. Lebofsky, AIP Conf. Proc. No. 83, p. 194 (1982).
13. D. Y. Gezari, R. Tresch-Fienberg, G. G. Fazio, W. F. Hoffmann, I Gatley, G. Lamb, P. Shu, C. McCreight, Ap. J., <u>299</u>, 1007 (1985).
14. I Gatley, T. J. Jones, A. R. Hyland, D. H. Beattie, T. J. Lee, M.N.R.A.S. <u>210</u>, 565 (9184).
15. I. Gatley, T. J. Jones, A. R. Hyland, R. Wade, T. R. Geballe, K. Krisciunas, M.N.R.A.S. <u>222</u>, 299 (1986).
16. D. N. B. Hall, S. G. Kleinmann, N. Z. Scoville, Ap. J. (Letters), <u>260</u>, L53 (1982).
17. T. R. Geballe, K. Krisciunas, T. J. Lee, I. Gatley, R. Wade, W. D. Duncan, R. Garden, E. E. Becklin, Ap. J., <u>284</u>, 118.
18. T. R. Geballe, this Symposium.
19. N. Kaifu, M. Hayashi, I. Gatley, this Symposium.
20. N. Z. Scoville, comment at this Symposium.
21. J. P. Henry, D. L. DePoy, E. E. Becklin, Ap. J. (Letters), <u>285</u>, L27 (1984).
22. W. J. Forrest, J. L. Pipher, W. A. Stein, Ap. J. (Letters), <u>301</u>, L49 (1986).
23. K. Y. Lo, this Symposium.

ATOMIC AND MOLECULAR GAS IN THE CIRCUMNUCLEAR DISK

R.Güsten

Max-Planck-Institut für Radioastronomie
Auf dem Hügel 69
5300 Bonn, F.R.Germany

ABSTRACT

The Sgr A West HII-region is surrounded by an extended (R≤8-9pc) massive ($M_d \approx 2-5\ 10^4 M_\odot$) disk of atomic and molecular gas. The structure's overall inclination to the line-of-sight is $\approx 70°$, and the major axis is tilted against the galactic plane. The thickness of the disk seems supported by turbulent pressure. The dominant large-scale motion is rotation ($v_{rot} \approx 110\ kms^{-1}$) about the nucleus. Recent high spatial resolution data reveal that the circumnuclear disk is not a planar equilibrium configuration, but is warped and kinematically perturbed on short timescales ($\approx 10^5 yr^{-1}$).

The atomic and molecular line emission arises from an unusually warm (T≈300K), dense ($n \approx 10^5 cm^{-3}$) and highly turbulent medium that occupies only a small volume of the emission region ($f_v \leq 0.1$). The excitation of the gas is due to a combination of UV heating, for the atomic phase, and dissipation of turbulent energy in small-scale low velocity shocks for the bulk of the molecular material.

The disk-like gas distribution may represent a parsec sized circumnuclear accretion disk, which in order to maintain the turbulence of the gas, must be in slow overall contraction towards the nucleus. The existence of the central cavity may require a recent ($\approx 10^5 yr$ ago) energetic nuclear event.

INTRODUCTION

Soon after the double-lobed appearance of the FIR dust emission towards Sgr A was determined[2], kinematic evidence for a ring of dust and gas orbiting around the galactic nucleus came from measurements of the [OI] $^3P_1 - ^3P_2$ (63 µm) fine-structure line on board the KAO[24,10]. In a series of papers Genzel et al.[11,12] (observing the atomic fine-structure transitions) and Liszt et al.[26,27] (studying atomic Hydrogen and molecular Carbon monoxide) investigated the global characteristics of the gas across the central few parsecs of the Galaxy in more detail. The data were interpreted in terms of a rapidly rotating ring of dense neutral material surrounding the Sgr A West HII-region. Since then, observations of numerous atomic [$O°, O^{++}, C^+$][11,12,30] and molecular [OH[35], CO[17,28,38], CS[23,5,16], NH_3[37,21], HCN[22,14] and HCO^+[34]] transitions, probing a wide range of excitation conditions, have given a fairly detailed picture of the physical conditions and

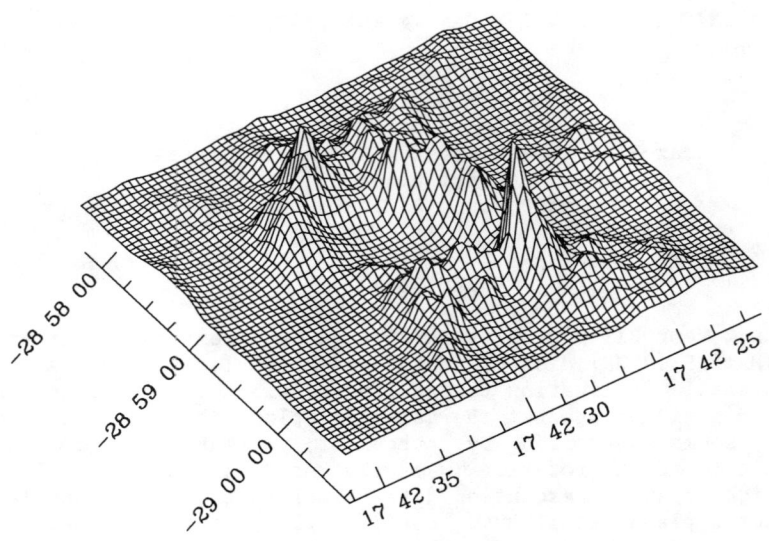

Fig.1 Relief representation of the velocity integrated HCN (J=1-0 88 GHz) emission, obtained with the UC Berkeley interferometer at Hat Creek[14]. The spatial resolution is 5-7".

energetics of the gas, its velocity field as potential probe of the central mass distribution, and the disk's relation to the nucleus.

GLOBAL CHARACTERISTICS

Fig.1 displays a relief representation of velocity integrated HCN(J=1-0,88.6GHz) emission obtained with the Hat Creek mm-wave interferometer of UC Berkeley[14]. This clumpy ring-like structure, which is elliptical in appearance, due to its inclination to the l-o-s, is to be interpreted as the sharp inner edge of an extended thin disk of a smoother density distribution that is resolved out in the synthesis data. In lower excitation, single dish [CII] 158μm[30] and CO(J=1-0[38]) and (2-1[28]) data, the full extent of the disk can be traced out to a galactocentric radius of 8-9pc. (Fig.2). The FWHM thickness of the structure (Fig.3) increases radially from $\approx$0.8pc at the inner edge to $\approx$2pc in the outer disk -roughly in hydrostatic equilibrium with the turbulent pressure of the gas[14]. The neutral gas is enveloping the bulk of the Sgr A West HII-region[2] (Fig.6), referred to as the 'central cavity' due to its significantly reduced gas density, of a few $10^2 cm^{-3}$ [2,12]. Their close correlation in velocity-space suggests an intimate physical relation between the ionized and neutral gas components. The 'northern arm', the 'eastern arm', and the 'western bar' ionized gas streamers appear physically coupled to the molecular ring[4,39,29,14]. The 'western arc' (R$\approx$1.7pc) is interpreted as the photo-ionized interface of the neutral disk that is exposed to the UV-field of the central illuminating source.

The overall inclination of the disk to the l-o-s is ≈70°. The major axis of the large-scale structure is tilted by ≈20° from the galactic plane. Model fits to the high resolution data obtained during the last few years revealed a complex, perturbed morphology of the disk. As a demonstration of the procedure, Fig.4 compares a model fit with an actual data strip, showing the change of velocity along an offset minor axis cut. The crescent shaped appearance is the expected pattern from an inclined rotating disk. The rotational and turbulent velocity components are directly inferred from the fits; the symmetry of the pattern limits any radial component to ≤20-30 kms^{-1}. The disk's intrinsic thickness is reflected in the latitudinal extent of the main emission rigde.

Fig.3 summarizes the major morphological characteristics of the circum-nuclear disk. One surprising finding is that the plane of rotation is changing with azimuth angle, suggesting the picture of a perturbed ('warped') non-equilibrium configuration with short

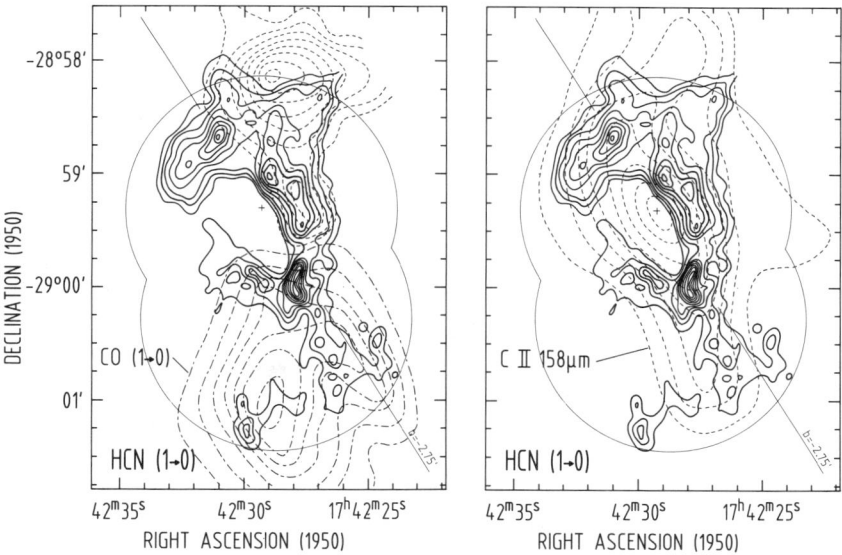

Fig.2 The circumnuclear disk as seen in (a) the lower excitation CO(1-0) emission {80<|v|<110 kms^{-1};beam:21"}[38], tracing the full extent of the structure, and (b) the [CII]158μm fine-structure line (beam:55")[30], arising from the UV-excited atomic gas phase. Superimposed is the HCN Hat Creek map (MEM processed, Fig.1), revealing the dense and clumpy nature of the inner disk. The more extended, smoother structure of the large-scale disk and the emission from l-o-s molecular clouds is resolved out in the synthesis data. This is different in single dish spectra, which especially towards the northern lobe suffer from blending with the "+50 kms^{-1} cloud".

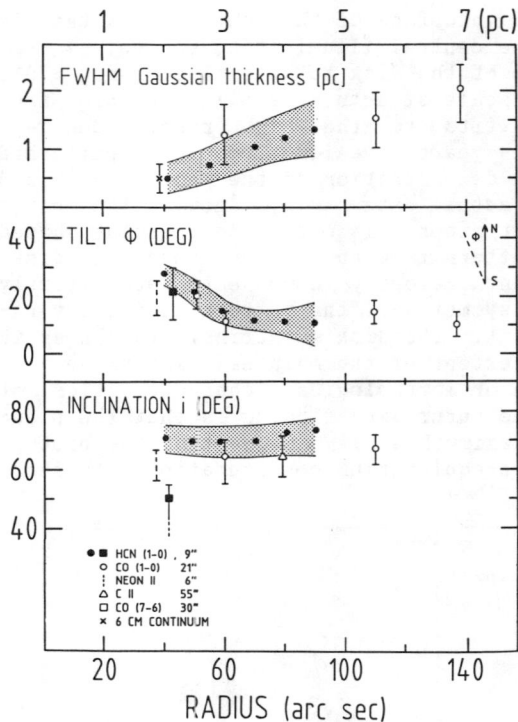

Fig.3 Physical characteristics of the neutral disk as derived from recent high resolution measurements. The FWHM thickness, the tilt angle (orientation of major axis on the sky) and the inclination versus radius are shown.[4,14,18,30,36,38]

timescales ($<10^5$yrs). While for the south-western (S-W) lobe an inclination of ≈70° is derived (90°:edge-on), the N-E (red-shifted) lobe appears to be tipped much stronger (≈50°, filled squares in Fig.3). The position angle ('tilt') of the major axis, as defined by the maximum observed velocity, changes systematically along the southern lobe by ≈20° towards N-S, with the inner edge oriented close to the galactic plane.

KINEMATICS

The velocity field has received considerable attention during the last years, because for a disk in equilibrium rotation, the inferred velocity field can be used to probe the central mass distribution[3]. Clearly, with observed peak velocities of ≈100 kms^{-1} rotation about the nucleus is the dominant large-scale velocity of the gas. But a controversial interpretation of the disk's velocity field concerns the kinematic evidence for a dominating central mass concentration (by Keplerian velocity fall-off). Harris et al.[17], on the basis of their CO(7-6) data along the galactic plane, and Lugten et al.[30], mapping the [CII] 158µm line, found a radially decreasing velocity of the 'main' emission ridge. This was taken as evidence that most of the mass enclosed within R=1.7pc is concentrated in a central point mass. Lo[28] and Serabyn et al.[38]

reached the opposite conclusion from their CO data. They quote a constant velocity with radius, consistent with rotation dominated by the gravitational field of the central stellar cluster. The apparent contradiction was elucidated only recently by higher resolution data, that revealed the complex and kinematically perturbed state of the structure in more detail[14]. Due to the warping plane of rotation, a derivation of the rotation curve from one data strip along a constant position angle (e.g. the galactic plane) will be misleading.

Corrected for the inclined and perturbed geometry, for the inner disk all data basically agree (Fig.5) and suggest a constant rotation velocity between $1.7 \leq R[pc] \leq 4-5$. If interpreted in terms of equilibrium orbits, the inferred gravitational field is consistent with the mass distribution of the central stellar cluster, $M_R \sim R^{1.2}$ [33], and constrains the mass of any central 'point' mass to $\leq 3\ 10^6\ M_\odot$. However, in view of the strong evidence for a non-equilibrium configuration (e.g. warping) of the disk, probing the mass distribution by its velocity field appears a doubtful procedure now[14]. For a more comprehensive recent discussion of the mass distribution across the central few parsecs of the Galaxy the reader is referred to references[13,39].

At larger radii the velocity pattern splits into two branches, one staying constant, the other bending off towards lower velocities (Fig.5). Whether this implies major changes in the velocity field, or can be also explained by a perturbed 'morphology'

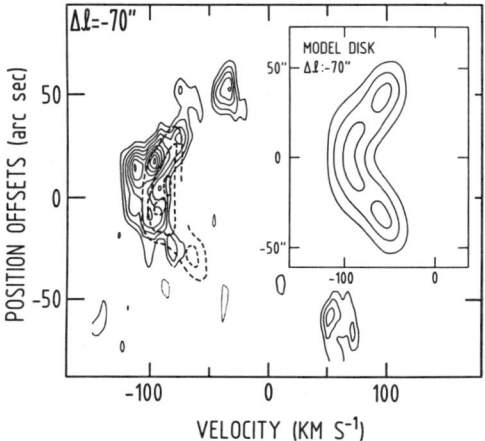

Fig.4 Composite HCN[14] and CO[38] position-velocity diagram, showing the change of velocity along an offset minor axis cut through the disk (offset: -70° along major axis). To take possible excitation gradients into account, the CO and HCN emission (probing lower and higher density material, resp.) are given superimposed. Model fits to the pattern (see insert) allow to determine the basic characteristics of the disk (thickness, orientation, velocities).

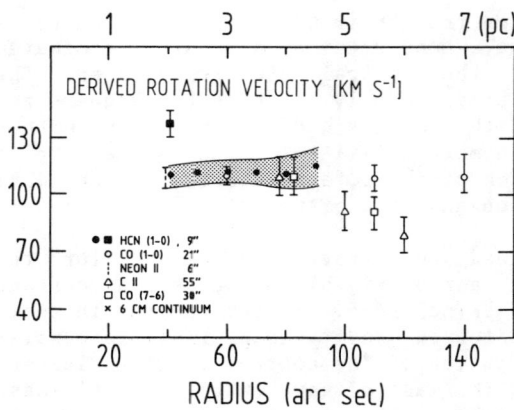

Fig.5 Derived rotation velocity versus g.c.radius, corrected for the perturbed morphology of the disk[14].

(e.g. azimuthally changing inclination and gradients in the excitation)[31,16] has to be studied in terms of an extended geometrical model as it has been developed for the inner disk.

On all scales, strong supersonic non-systematic motion is superimposed on regular rotation. Unresolved small-scale 'turbulence' as high as 50-70 km s^{-1} is present in most of the prominent HCN clumps (size ≤0.5 pc), yielding a dynamical age of ≤10^{4}yrs only! Also the velocity scattering between these eddies is significant. The most extreme example is the prominent feature close to the western arm, where the bar crosses the ring, which velocity differs from that expected for an uniformly rotating disk by ~50 km s^{-1}. An important finding of the high-resolution HCN synthesis data is that the feature is resolved from the underlying smoother ring emission, and is identified as an 'extra feature', not representative for the bulk motion of the ring (Güsten et al., their Fig.2).A global radial velocity component of ~50 km s^{-1} for the inner disk[8,23,34] is not supported by these data. On the contrary, the above mentioned symmetry of the velocity structure along minor axis cuts (Fig.4) limits any radial velocity component to ≤20-30 km s^{-1}.

PHYSICAL CONDITIONS

Observations of a variety of atomic and molecular transitions, probing a wide range of excitation conditions, have revealed that the emission of both collisionally and UV-excited species is coexistent throughout the disk. Fig.2 displays the spatial overlap between the ionized atomic [C$^+$] and neutral molecular [CO] emission. Obviously, soft UV radiation is penetrating deep into the disk. The excitation conditions for the atomic and molecular emission are similar, both arising from a fairly dense warm medium.

Towards the inner disk from a number of tracers, Hydrogen <u>densities</u> in the emission region of n(H,H$_2$)=10$^{5\pm.5}$cm^{-3} are derived[12,17,34,5,16], that are substantially higher than the volume averaged gas density, ⟨n⟩=10$^{3.6\pm.3}$cm^{-3}. The latter is inferred from the gas column density N(H,H$_2$)≈4(±2) 10^{22} cm^{-2}, as estimated from the dust continuum emission[12] and the CO line emission[38], and an assumed scale-length of ≈3pc. The implied overall volume filling factor is small, $f_V = \langle n \rangle / n \approx 10^{-1.4 \pm .8}$. There is some evidence that the warm less dense ($10^{4.6}$ cm^{-3}) CO emitting gas[31] fills a larger fraction of the volume, $f_V^{CO} \approx 0.1$, than does the higher density ($10^{5.3}$ cm^{-3}) CS exciting gas[5,16], $f_V^{CS} < .05$. Those resolved structures that can be identified in the higher resolution synthesis maps (Figs.2,6; see also [21,35]) represent the largest eddies only in a wider spectrum of density fluctuations.

Recent CO[31] and CS[16] excitation studies indicate that the density decreases radially by a factor ≥5 across the disk. But nowhere in the disk gas densities comparable to Roche critical densities for tidal stability, $n_c \approx 10^7$ cm^{-3} at R~2pc, are observed. Hence tidal stripping (in addition to inter-clump collisions) is likely the cause of the enormous intraclump velocity dispersion and their short dynamical lifetime.

The total **mass** of the disk as estimated from the dust continuum emission[12] and CO measurements[38] amounts to $M_d \approx 2\text{-}4 \cdot 10^4 M_\odot$. This is ≈$10^3$ times the mass of the ionized gas in the central cavity, but represents only a very small fraction, 10^{-3}, of the total matter locked-up in the form of stars. The mass of atomic Hydrogen is only ≈$5 M_\odot$ ($T_S/15K$) [27], revealing the overwhelming molecular nature of the matter.

Multi-line CO studies[17,18,31], covering a wide range of J-transitions yield 'molecular gas' **temperatures** $T_g \approx 300\text{-}400$ K towards the inner edge of the disk. For the atomic phase very similar (≈350K)[12], from the more fragile NH$_3$ molecule somewhat lower temperatures (≈250K)[37] have been derived. The 'CO'-temperatures seem to decrease moderately with distance from the nucleus[31]. Bordering the ionized interface, a thin layer ($N \approx 10^{17}$ cm^{-2}) of hot ($T \geq 10^3$) vibrationally excited molecular Hydrogen is observed[7,8]. Their spatial and kinematical correspondence suggests that the hot gas is physically coupled to the warm molecular material. The intensity ratio between the v=2-1 to v=1-0 S(1) H$_2$ transition is suggestive of shock excitation, possibly due to interaction of interface gas with the high-velocity wind associated with the galactic nucleus[9].

However, the energy carried with the wind falls short by orders of magnitude to account for the total luminosity $L_T \approx 2\text{-}5 \cdot 10^4 L_\odot$ of the atomic and molecular lines from all the disk. An important clue for the discussion of alternative gas **heating mechanisms** is that the temperature of the dust particles, $T_c \approx 40\text{-}60K$ [2,25] is much below the kinetic gas temperature. $T_g \gg T_c$ excludes collisions with radiatively heated warm dust particles as the dominant heating process, and requires a mechanism that acts directly on the gas.

Most likely, the excitation of the gas is a combination of UV- and slow shock heating. Soft (<13.6eV) photons may excite vibrational states in the H$_2$ molecule and/or may eject 'hot' electrons from grain surfaces[40,11], which energy is thermalized then by collisions with ground-state H$_2$ molecules. From the intensities of the atomic fine-structure lines, UV-heating seems especially attractive for the atomic gas phase[12]. Due to the high clumpiness of the gas, the UV field can penetrate deep into the disk and create thin photodissociation layers at the surfaces of exposed cloudlets. The stronger concentration of atomic ('photodissociated') material towards the inner disk is explained by the gradually increasing attenuation of the UV field. However, the high temperature of the molecular phase is difficult to reconcile with UV heating[37,17], and requires some alternative heating process. First, photoelectric heating is effective ($T_g \gg T_d$) only up to a penetration depth of ≈ 3 mag[40], while the column density across the largest HCN eddies corresponds to 10-100 mag, and second, the same UV photons that heat the electron gas will also photodissociate the fragile more complex molecules (to some degree only, CO with its high ionisation potential (11eV) may survive due to self-shielding).

As for the molecular clouds on much larger scales[15], dissipation of turbulent energy in small-scale shocks may also account for the heating of the molecular disk gas. With the high turbulence that is characteristic for the disk on all scales, $E_{turb} \approx 5 \; 10^{50}$erg, turbulent heating requires small efficiencies only[37,17], but has to be maintained continuously on a timescale of $\approx 10^5$yrs. For this, the gravitational energy of the disk, stored in the differential rotation, may be tapped[6], resulting in a slow overall contraction of the disk towards the nucleus[14]. To balance the observed cooling luminosity of the gas, $L_T \approx 2-5 \; 10^4 L_\odot$, a net mass inflow rate of $\approx 5 \; 10^{-2} M_\odot yr^{-1}$ is required.

GENESIS OF THE DISK

The question about the origin and evolution of the circumnuclear disk is likely directly related to the presence of the central cavity. In the above developed picture of a slowly contracting large-scale accretion disk, the dynamical timescale of the cavity is $\approx 10^3$yr only. On a similar timescale the sharp interface layer canot be in dynamical equilibrium, as neither radiative pressure, thermal pressure of the HII-region nor the ram pressure of the nuclear wind can counterbalance the large ram pressure of the turbulent disk ($\approx 10^{-(6-7)}$dyn cm^{-2}). Some energetic agent seems required, that cleared the cavity some $\approx 10^5$yr ago, which may have been explosive ($\geq 10^{51}$erg) or steady mass loss with an average rate $\geq 0.2 \; M_\odot yr^{-1} v 500$ kms^{-1}. The present configuration would then represent a transient phase, in which material is slowly accreting towards the nucleus again, and which may finally lead to anew, recurrent energetic event (star burst, 'explosive' accretion towards a possible black hole).

Alternatively it has been suggested that the disk reflects the strong dynamical resonance of the gas to a non-axisymmetric perturbation of the gravitational field[12]. However, this model seems to require stronger radial velocity components than observed in the new data. In their recent magneto-hydrodynamical model Heyvaerts et al.[19] argue for a gradual built-up of the molecular

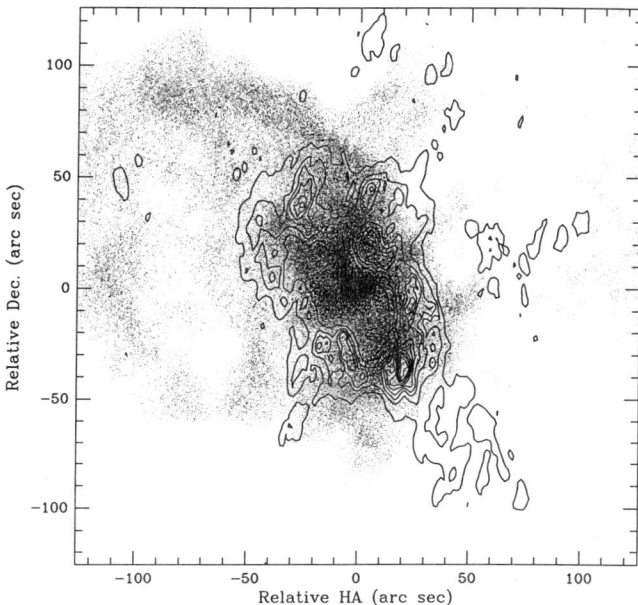

Fig.6 High resolution HCN map, superimposed on a grey-scale representation of the 6cm continuum emission[4]. The HCN map is from the MEM processed complete Hat Creek data set of 45 baselines, yielding 2-3" resolution[41]. A close morphological correspondence between the dense molecular gas and the thermal 'spiral' as well as the non-thermal Sgr A West 'diffuse' component is obvious. Note that several of the non-thermal protrusians appear to extend beyond the otherwise well-defined interface layer.

disk from mass-carrying expanding magnetic loops, ejected from the central engine. The existence of non-thermal protrusions which morphologically interact with the inner disk (Fig.6) and the non-thermal nature of the diffuse Sgr A West component[4], may fit into their scenario. However, both models face the difficulty that, the clearing agent for the cavity, and the process that maintains the overall turbulence in the disk, have to be addressed separately. If the ring should be kept confined by magnetic pressure, a high magnetic field strength of 5-10 mG has to be required. Although for the disk there are no measurements available, interestingly towards the 'northern arm' from dust polarisation measurements a field strength of this magnitude has been derived recently[1].

The origin of the disk gas is a matter of discussion. As an alternative to nuclear expulsion[32,19], the material may have been tidally stripped from nearby more massive molecular clouds[21]. Also diffuse lower angular momentum gas, released from evolving stars of the old cluster, may be accumulating close to the nucleus within $\approx 10^{7\pm 1}$ yr. Chemistry sensitive observations may be able to discriminate between these possibilities.

REFERENCES

1. Aitken,D.K., Roche,P.F., Bailey,J.A., Briggs,G.P., Hough,J.H., Thomas,J.A.: 1986, MNRAS 218,363
2. Becklin,E.E., Gatley,I., Werner,M.W.: 1982, Ap.J. 258,123
3. Crawford,M.K., Genzel,R., Harris,A.I., Jaffe,D.T., Lacy,J.H., Lugten,J.B., Serabyn,E., Townes,C.H.: 1985, Nature 315,467
4. Ekers,R.D., van Gorkom,J.H., Schwarz,U.J., Goss,W.M.: 1983, Astr.Ap. 122,143
5. Evans,N.J. II: 1986, paper contributed to this conference
6. Fleck,R.C.: 1980, Ap.J. 242,1019
7. Gatley,I., Jones,T.J., Hyland,A.R., Beattie,D.H., Lee,T.J.: 1984, MNRAS 210,565
8. Gatley,I., Jones,T.J., Hyland,A.R., Wade,R., Geballe,T.R., Krisciunas,K.L.: 1986, MNRAS 222,299
9. Geballe,T.R.: 1986, review paper from this conference
10. Genzel,R., Watson,D.M., Townes,C.H., Lester,D.F., Dinerstein,H., Werner,M.W., Storey,J.W.V.: 1982, in "The Galactic Center", AIP Conf.No.83, eds. G.R.Riegler & R.D.Blandford
11. Genzel,R., Watson,D.M., Townes,C.H., Dinerstein,H.L., Hollenbach,D., Lester,D.F., Werner,M.W., Storey,J.W.V.: 1984, Ap.J. 276,551
12. Genzel,R., Watson,D.M., Crawford,M.K., Townes,C.H.: 1985, Ap.J. 297,766
13. Genzel,R.: 1987, Proc.1986 NATO Summer School "The Galaxy", Cambridge, eds.G.Gilmore & R.Carswell, Reidel
14. Güsten,R., Genzel,R., Wright,M.C.H., Jaffe,D.T., Stutzki,J., Harris,A.I.: 1987, Ap.J. in press
15. Güsten,R., Walmsley,C.M., Ungerechts,H., Churchwell,E.: 1985 Astr.Ap. 142,381
16. Güsten,R., Serabyn,E., Evans,N.J.II, Downes,D.: 1987, in prep.
17. Harris,A.I.: 1986, Ph.D. Thesis, University of Calif.,Berkeley
18. Harris,A.I., Jaffe,D.T., Silber,M., Genzel,R.: 1985, Ap.J.(Letters) 294,L93
19. Heyvaerts,J., Pudritz,R.E., Norman,C.A.: 1986, paper contributed to this conference
20. Ho,P.T.P., Jackson,J.M., Barrett,A.H., Armstrong,J.T.: 1985 Ap.J. 288,575
21. Ho,P.T.P., Jackson,J.M., Barrett,A.H.: 1986, paper contributed to this conference
22. Kaifu,N., Inatani,J., Hasegawa,T., Morimoto,M.: 1983, in IAU Synp.106 "The Milky Way Galaxy", eds. H.van Woerden, W.B.Burton & R.J.Allen

23. Kaifu,N., Hayashi,M., Gatley,I.:1986, paper contributed to this conference
24. Lester,D.F., Werner,M.W.,Storey,J.W.V.,Watson,D.M.,Townes,C.H.: 1981, Ap.J.(Letters) 248,L109
25. Lester,D.F., Joy,M., Harvey,P.M.,Ellis,Jr.H.B.: 1986, paper contributed to this conference
26. Liszt,H.S., van der Hulst,J.M., Burton,W.B., Ondrechen,M.P.: 1983, Astr.Ap. 126,341
27. Liszt,H.S., Burton,W.B., van der Hulst,J.M.: 1985, Astr.Ap. 142,237
28. Lo.K.Y.: 1986, Science 233,1394
29. Lo.K.Y., Claussen,M.J.: 1983, Nature 306,647
30. Lugten,J.B., Genzel,R., Crawford,M.K., Townes,C.H.: 1986, Ap.J. 306,591
31. Lugten,J.B., Harris,A.I., Storey,G.J., Genzel,R., Townes,C.H.: 1986, paper contributed to this conference
32. Rees,M.J.: 1986, review paper from this conference
33. Sanders,R.H., Lowinger,T.: 1972, Astr.J. 77,292
34. Sandqvist,Aa., Wooten,A., Loren,R.B.: 1985, Astr.Ap. 152,L25
35. Sandqvist,Aa., Karlsson,R., Whiteoak,J.B., Gardner,F.F.: 1986, paper contributed to this conference
36. Serabyn,E., Lacy,J.H.: 1985, Ap.J. 293,445
37. Serabyn,E., Güsten,R.: 1986, Astr.Ap. 161,334
38. Serabyn,E., Güsten,R., Walmsley,C.M., Wink,J., Zylka,R.: 1986, Astr.Ap. 169,85
39. Serabyn,E., Lacy,J.H., Townes,C.H.: 1987, preprint
40. Thielens,A.G.G.M., Hollenbach,D.: 1985, Ap.J. 291, 722
41. Wright,M.C.H., Genzel,R., Güsten,R.,Plambeck,R.L.,Carlstrom,J.: 1987, in preparation

THE GALACTIC CENTER COMPACT NONTHERMAL RADIO SOURCE

K. Y. Lo

Astronomy Department
University of Illinois, Urbana-Champaign
and
California Institute of Technology

ABSTRACT

The current observational status of Sgr A*, the compact nonthermal radio source at the galactic center, is reviewed. It is a unique radio source at a unique position of the Galaxy. It is unlike any compact radio sources associated with known stellar objects, but it is similar to extragalactic nuclear compact radio sources. The positional offset between Sgr A* and IRS16 places little constraint on the nature of the underlying energy source, since the nature of IRS16 itself is not well understood and may not be the core of the central star cluster. With its unique properties in the Galaxy and being the only unusual object at the center with dimensions approaching the gravitational radius of a $\sim 10^6$ $M_\odot$ black-hole, Sgr A* is still the best candidate for marking the location of a massive collapsed object.

INTRODUCTION

At the center of Sgr A, the powerful radio source at the center of the Galaxy, is an extremely compact nonthermal radio source.[1,2] Such a radio source was anticipated as a possible signature of a massive black-hole that was proposed to be the energy source for the luminosities observed towards the center.[3]

Efforts to determine the structure of this compact radio source have gone on for many years,[4-6] but because of the practical difficulties there is still no brightness distribution map of the source. The principal limitation, given the existing radio telescopes and the angular scale of the source, is that there are very few sufficiently short baselines on which the source is not completely resolved.

We review here the current status of the properties of the radio source and the constraints one can place on the nature of the underlying energy source. We also discuss briefly the relevance of various other constraints that have been placed on the source.

THE COMPACT NONTHERMAL RADIO SOURCE

The properties of the radio source is summarized in Table 1.

Table 1 Observed properties of the galactic centre compact radio source

Source size	< 16 AU ($\sim 2 \times 10^{14}$ cm)
Wavelength dependence of size	$\lambda^{2.0}$ ($\lambda \geq 1.35$ cm)
Position angle of elongation	$98 \pm 15°$ ($\lambda = 3.6$ cm)
Axial ratio	0.55 ± 0.25 ($\lambda = 3.6$ cm)
Upper limit to source expansion	< 13 km s^{-1}
Flux density variabiality ($\Delta S/2S$)*	$\lesssim 0.2$ (1975-86)
Spectral index	~ 0.25
Turnover frequency	$\gtrsim 89$ GHz
Radio luminosity	$\sim 1.3 \times 10^{34}$ erg s^{-1}
Brightness temperature	$> 7 \times 10^8$ K (λ 1.35 cm)

*ΔS, peak-to-peak variation; S, mean flux density; ref. 15 and J. van Gorkom, K. Y. Lo and M. Claussen, unpublished results.

Scale Size

Interferometric and VLBI observations have yielded source diameters (FWHM of a Gaussian brightness distribution, the simplest source model) of Sgr A* at various wavelengths.[2,6,7,8] The results are summarized in Figure 1. The smallest measured size is 0.002" ± 0.0003", corresponding to 16 AU at a distance of 8 kpc. The diameters are seen to vary very nearly as λ^{-2}. The formal value of the index derived from the measurements as summarized in figure 1 is 2.08 ± 0.05. Given the systematic errors involved and the lack of proper maps, the index cannot be distinquished from 2.

Davies et al[7] first suggested that the apparent source size of Sgr A* could be affected by *interstellar scattering* of the radio wave, due to the irregular distribution of interstellar electrons along the line of sight. In this case, the observed size is an upper limit to the intrinsic size of the source. However, the amount of scattering is among the highest observed and the amount of electron density fluctuations required depends on where the bulk of scattering occurs.[9]

There are indications that sources within 100 pc of the center have large observed sizes implying a source of strong scattering near the center,[10] but Cordes et al[11] suggest that the observed scattering in Sgr A* is not an unrealistic extrapolation from pulsar results which can be explained by a two-component model of general interstellar electrons. The reasons for where to place the blame of scattering are twofold – one has to do with whether to believe scattering is affecting Sgr A* because a λ^{-2} dependence of the source size could be due to a 1/r distribution of thermal electrons within the source,[12] and two is the possible variability time scales involved, due to scintillation and refractive scattering.[13] In any case, the current measurements are only upper limits to the source size.

Source Shape

Recent VLBI observations have revealed structure more complicated that the simplest model of a symmetrical Gaussian distribution.[2] At 3.6 cm, the observations are best fit by an elliptical Gaussian, with an axial ratio of 0.55 ± 0.25. The position angle of the major axis is 98 ± 15 degrees, whereas the minor axis of the Galaxy is at 122 degrees. Due to the limited amount of data, the observed elongation in the source structure should be verified with more extensive observations. Clearly, brightness distribution maps of Sgr A* would help to determine the nature of the radio source. The best way to obtain them is at high frequency where both propagation effects and the opacity due to thermal electrons would be minimized.

Many issues involved in the effects of propagation through the interstellar electrons on the observations of Sgr A* are not definitively resolved, and more observations are needed. For example, if the radio waves from Sgr A* are strongly scattered, the observed image is unlikely to be full of details. Refractive scattering may also shift the positions of the radio source with time as well as with wavelength.[14] The presence of intensity scintillation in Sgr A*, if observed, would place a very strong constraint on the source size.[9]

Variability

Since Sgr A* was identified in 1974, there have been many measurements of its flux density by various observers. The observed flux density was often different but never varying by more than ∼ 20 percent ($\Delta S/2S$). The only published study of the variability of Sgr A* was that of Brown and Lo[15] between December 1976 and May 1978. They found a few instances of low level variation on hourly time scale as well as a secular increase at 11 cm. The secular increase at 11 cm has not continued indefinitely since later measurements at nearby wavelengths have not shown an unusually large value. The variability properties of Sgr A* are not very well defined, except that it is clear that large outburst is not common and has not been observed.

A multi-wavelength study of variability at different time scales may reveal intrinsic source variability, or changes due to scintillation[9] and refractive scattering effects,[13] all of which could help to resolve the important issues of source sizes, propagation effects and radiation mechanisms. This kind of study requires dedicated, carefully calibrated and sensitive interferometer observations.

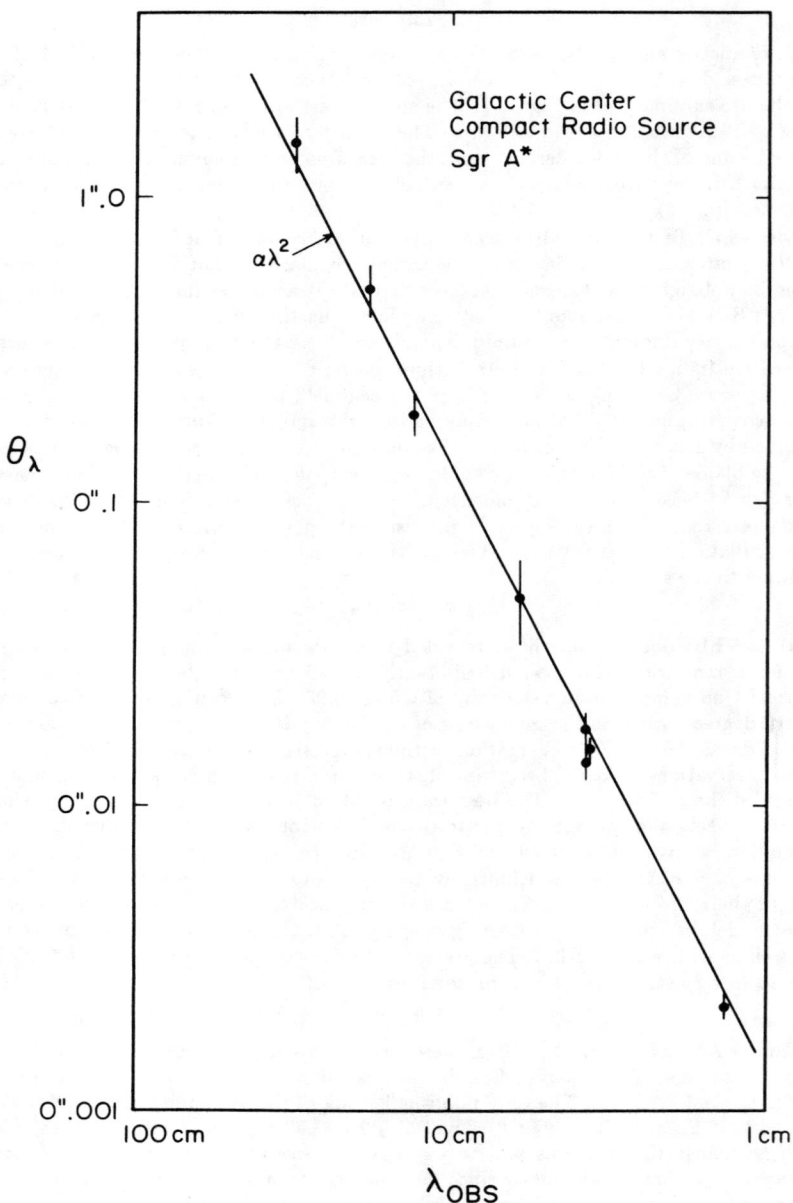

Figure 1. A plot of measured source diameter versus observing wavelength.[4-8]

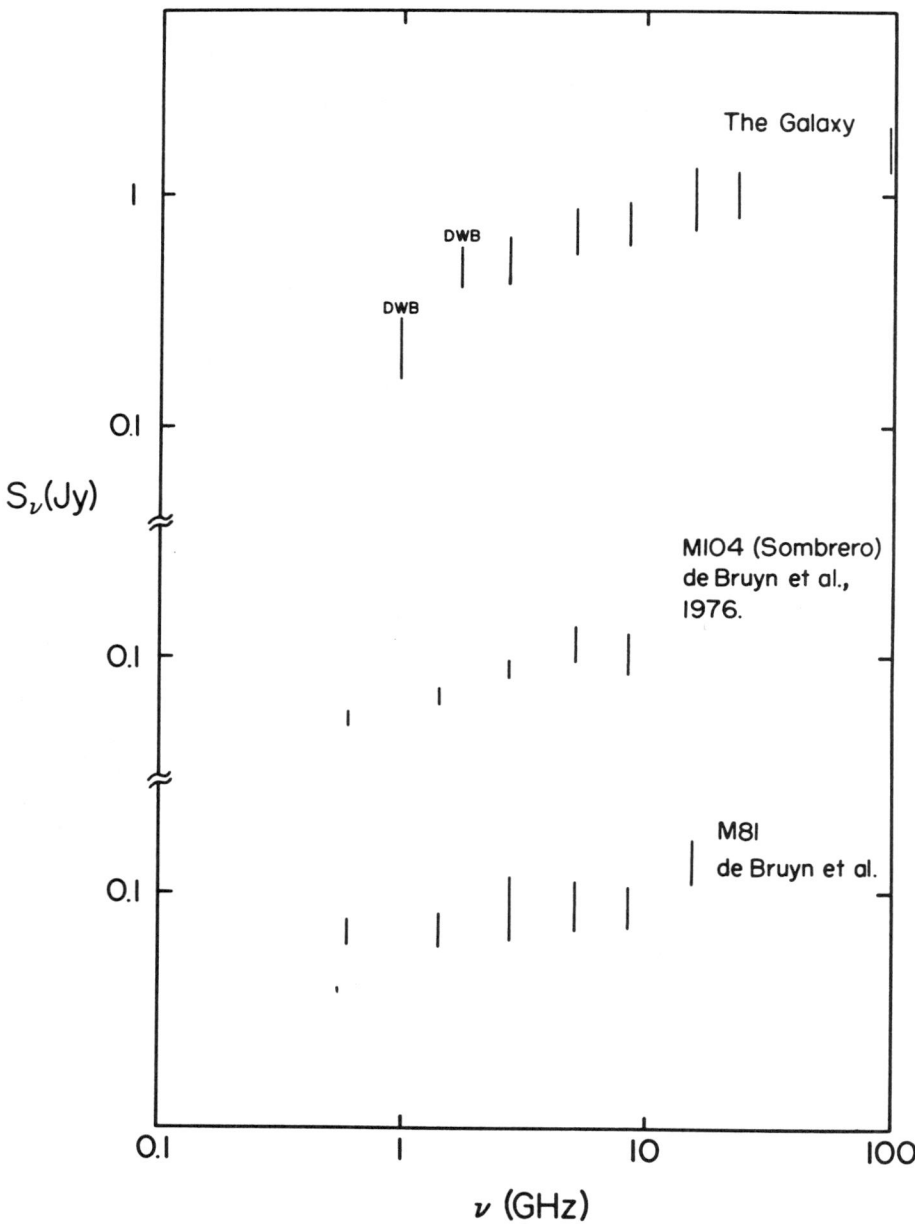

Figure 2. A log-log plot of flux density ($Wm^{-2} Hz^{-1}$) versus observing frequency for the compact nonthermal radio sources at the center of the Galaxy, M81 and M104 (the Sombrero Galaxy).

Source Spectrum

The spectrum of a radio source may also provide information on the source structure.[16] Figure 2 shows a "mean" spectrum of Sgr A* along with the spectra of the compact radio sources in the nuclei of M81 and M104.[17] The slowly rising spectra are most simply interpreted as that of a self-absorbed incoherent synchrotron source. The highest frequency measurement is at $\sim$ 90 GHz where the flux density is 1.05 $\pm$ 0.15 Jy.[18] The turn-over frequency, above which the source becomes optically thin, is thus > 90 GHz, implying a very compact source. An important empirical point is that the source spectrum of Sgr A* is completely unlike that of pulsars.

UNDERLYING ENERGY SOURCE

Even if the brightness distribution of Sgr A* is as yet unknown, its very small scale size poses a very strong constraint on the underlying energy source and requires objects with stellar dimensions. In the Galaxy, there are very compact radio sources associated with known stellar objects - pulsars, binary stellar radio sources, young supernovae or supernova remnants. However, empirical arguments can already exclude all of these as possible energy source.

Ordinary pulsars are ruled out because the spectral index of pulsar radiation, α, generally lies in the range $-3 < \alpha < -1$, whereas that of Sgr A* is $\sim$0.2 and the most luminous pulsars have an average radio luminoisty of about 10^4 times less. Furthermore, pulsars are very common, whereas Sgr A* have unique properties in the Galaxy. Binary stellar radio sources are unlikely to be the energy source of Sgr A* because they are characterized by frequent, large outbursts in which $\Delta S/2S$ could be 10 to 1000, while their steady radio luminosity is typically 10^5 times smaller that of Sgr A*. The ratio of their X-ray to radio luminosities is least 30 times larger than that of Sgr A*.

Recently, there have been detections of very bright compact nonthermal radio sources in external galaxies that are most likely young supernovae and supernova remnants.[19] They are different from Sgr A* in that they have been observed to expand or their flux densities are declining steadily on the time scale of a few years. For more than 12 years, Sgr A* have been essentially steady in its intensity and the expansion velocity is constrained to be <13 km/s averaged over 8 years.[2]

While one cannot explain Sgr A* in terms of known stellar radio sources, the radio luminosity of Sgr A* is only $\sim 10^{34}$ erg s^{-1} or $\sim$ 10 L$_\odot$. As up to 10^{38} erg s^{-1} may be available from a rotating neutron star, at least in principle, Sgr A* can be powered by a rotating neutron star. A model of pulsar-driven wind has been proposed for Sgr A*,[16] but the small scale size and the lack of dense gas clouds at the position of Sgr A* (which is needed to pressure confine the source) have ruled out such a model.

The mass of Sgr A* is thus a critical parameter needed to determine the nature of the energy source. There is an indirect way of constraining the mass – its proper motion. If Sgr A* is a stellar mass object near the bottom of the potential well at the center, very likely it will possess a large velocity. Hence, a measurable proper motion would rule out the possibility that it is supermassive. The latest proper motion result shows that the motion of the sun is observed relatively to Sgr A* and not relative to the neighbouring reference sources and that the residual proper motion is less than 40 km s^{-1}.[20] Thus, Sgr A* is at the galactic center, and the low upper limit to the proper motion, while not a proof, is consistent with Sgr A* being supermassive.

SIMILARITY TO EXTRAGALACTIC COMPACT RADIO SOURCES

While Sgr A* is a unique radio source in a unique position in the Galaxy, it has properties very similar to those of extragalactic compact radio sources. In particular, in the nucleus of M81, a nearby bright spiral (Sb) galaxy, there exists such a radio source which has been quite well observed.[21] It is compact (1000 AU $\times$ 4000 AU), with the major axis essentially aligned with the minor axis of the galaxy. It has a rising spectrum (cf. figure 2) and is mildly variable[17]. It is also a unique radio source at the center of a bright galaxy.[22] However, in this case, the radio luminosity ($\sim 10^{37}$ erg s^{-1}) is high enough that powering by a rotating neutron star is unlikely. Rees et al have proposed that extragalactic radio sources that are not luminous in the optical are explainable by massive black holes with a low accretion rate.[23] In particular, the model has been applied to Sgr A*.[24]

IDENTIFICATION OF SGR A*

The relatively low luminosity of Sgr A* at radio wavelenghts ($\sim 10^{34}$ erg s^{-1}) makes it still possible for a rotating neutron star to power it. Identification of Sgr A* with objects in the infrared is thus important for determining its bolometric luminosity. For example, a total luminosity of 10^7 $L_\odot$ is inferred to originate from the central pc of the Galaxy.[25] Whether all this luminosity is produced in a single object or a cluster of young stars has not been resolved.[26,27] A large bolometric luminosity would place a further important empirical constraint on the underlying energy source. Identification would also be helpful for identifying the location of Sgr A* relative to the central star cluster and for dynamical probes of the mass of the radio source.

The identification process has been beset by systematic errors involved in determining absolute positions of the near-infrared sources.[26,28] In the radio, the best absolute position of Sgr A* has been determined with the VLA,[29] but the atmospheric modelling sets the limiting systematic errors of ± 0.2". Ekers was the first to suggest a method which bypassed many of the systematic errors involved in comparing the absolute positions of the radio source and the near-IR sources.[30] He compared the Brγ (2.16 μm) map to a 6 cm radio map. By aligning the ionized gas features observable at both 2 μm and 6 cm, he concluded that Sgr A* is offset from IRS16, a 2 μm continuum source. The best result using this method is that of Forrest et al who obtained a Brα (4.05 μm) image with an 1 - 5 μm array camera.[31]

Thus, the evidence seems to point to a positional offset between Sgr A* and the various components of IRS16.[30,28,31] However, since the nature of the IRS16 components is uncertain,[32] the implication of this "non-identification" is not clear. The non-association of an obvious near-infrared source with Sgr A* does not preclude the possibility of a massive black-hole, since in the case of low accretion rate onto a massive black-hole, which could explain Sgr A*, little luminosity is expected from the ion-supported accretion disk.[23] In addition, the actual accretion process is highly uncertain, Sgr A* may be in a "starved" phase.[24]

If Sgr A* were a $10^6 M_\odot$ massive black-hole, a cusp of stars is expected to form around it with an accretion radius of 20".[33,34] The 2 μm light distribution within the central few arcseconds is complicated by a few bright discrete sources, so that the determination of the peak of the light distribution is not straight-forward.[35] High resolution measurements of stellar velocity dispersion may be the most reliable way to define the dynamical center of the Galaxy.[36,37] Even then, uncertainties in the intrinsic velocity distributions may still lead to ambiguous results.[38]

UPPER LIMITS TO MASS OF SGR A* ?

There have been various discussion on the mass of the possible black hole at the galactic center based on the separation of Sgr A* from IRS16,[39,34] the broad He line[40,41] and positron production[42,41]. It has been argued that the mass of the possible black hole cannot be large.[41,34] But, given the uncertainties in the nature of the components of IRS16, the location of the dynamic center, the ill-defined location of the 0.5 Mev e^+e^- annihilation line and alternate ways to produce positrons,[23,43] the arguments are not definitive.

The most direct observational indication of the possible presence of a concentrated mass at the center is still given by the large velocities of the ionized gas.[44] Given the complexities of the ionized gas distribution (figure 3) and the uncertain origin of its motion, the gas velocities may not accurately probe the gravitational potential. The origin and motion of the ionized gas will be much better constrained if the three dimensional velocity field of the gas is known. We

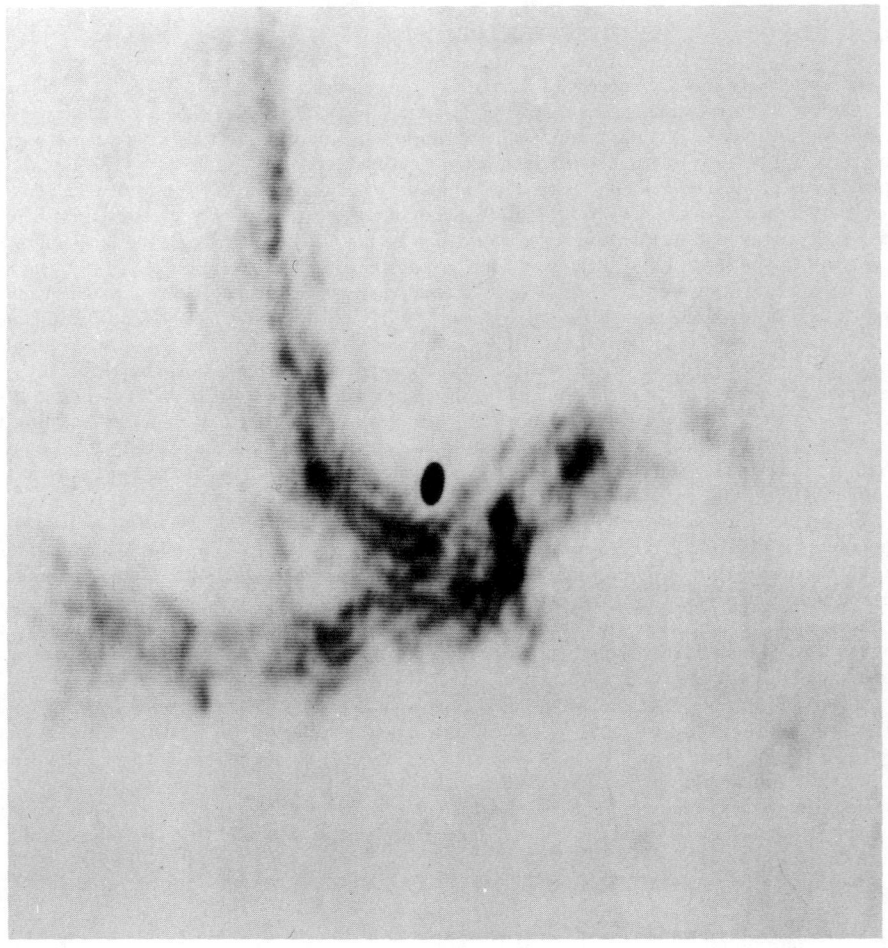

Figure 3. A 0."3 x 0."6 resolution λ6cm radio continuum VLA map of the central 45" of SgrA West.[45] The ellipse is the "saturated image" of SgrA* and is much bigger than the actual size. Note the filamentary structure in the arms and suggestive braided structure in the filaments along the northern arm.

plan to measure the proper motion of the ionized gas by obtaining another high resolution map with the VLA in 1987, for comparison with the map in figure 3 obtained in 1983. A transverse motion of 100 km s^{-1} corresponds to a proper motion of 0.01" in this period. This proper motion measurement can only improve with time as the time base increases.

CONCLUSION

Accretion onto black-holes was proposed to account for the inordinately large luminosity observed from a very compact volume of space in quasars and radio galaxies. Observationally, there is almost no hope of spatially resolving such collapsed objects in such distant objects. In the center of our Galaxy, the proximity permits probing within the interesting scales approaching the Schwarzschild radius of a 10^6 $M_\odot$ black-hole, as in the case of Sgr A*. Ironically, while the unique properties of Sgr A* preclude models involving known stellar systems, the radio luminosity is too low to demand the mechanism of accretion onto a massive black hole. The exact nature of Sgr A* can only be resolved with more observations.

REFERENCES

1. Balick, B., and Brown, R. L., *Astrophys.J.*, **194**, 265-270 (1974).
2. Lo, K. Y., *et al, Nature*, **315**, 124-126 (1985).
3. Lynden-Bell, D., and Rees, M. J., *Mon. Not. R. Astr. Soc.*, **152**, 461-475 (1971).
4. Lo, K.Y., Schilizzi, R. T., Cohen, M. H., Ross, H. N., *Astrophys. J.*, **218**, 668-670 (1977).
5. Kellermann, K. Y., Shaffer, D. B., Clark, B. G., and Geldzahler, B. G., *Astrophys.J.Lett.*, **214**, L61-L62 (1977).
6. Lo, K. Y., Cohen, M. H., Readhead, A. C. S., and Backer, D. C., *Astrophys.J.*, **249**, 504-512 (1981).
7. Davies, R. D., Walsh, D., and Booth, R., *Mon. Not. R. astr. Soc.*, **177**, 319-333 (1976).
8. Marcaide, J. M., et al., in *Conference on Active Galactic Nuclei* (ed. Dyson, J. E.) (Manchester University Press) 50-53 (1985).
9. Ozernoi, L. M., Shishov, V. I., *Sov. Astron. Lett.*, **3**, 233-235 (1977).
10. Backer, D. B., *Astrophys.J.Lett.*, **222**, L9-L12 (1978).
11. Cordes, J. M., Ananthakrishnan, S., and Dennison, B., *Nature*, **309**, 689-690 (1984).
12. de Bruyn, A. G., *et al, Astron. & Astrophys.*, **46**, 243 (1976).
13. Rickett, B. J., *Astrophys.J.*, **307**, 564-574 (1986).
14. Romani, R. W., Narayan, R., Blandford, R. B., *Astrophys. J.*, in press.
15. Brown, R. L., Lo, K. Y., *Astrophys.J.*, **253**, 108-114 (1982).
16. Reynolds, S. P., and McKee, C. F., *Astrophys.J.*, **239**, 893-897 (1980).
17. de Bruyn, A. G., *Astron. & Astrophys.*, **52**, 439 (1976).
18. Wright, M. C. H., this volume.
19. Kronberg, P. P., Sramek, R. A., *Science*, **227**, 28 (1985).
20. Backer, D. B., Sramek, R. A., this volume.
21. Bartel, N., *et al. Astrophys.J.*, **262**, 556-563 (1982).
22. Bash, F., Kaufman, M., *Astrophys.J.*, **310**, 621-636 (1986).
23. Rees, M., *et al Nature*, **295**, 17-20 (1982).
24. Rees, M., *AIP Conf. Proc.*, **83**, 166-176 (1982).
25. Becklin, E., Gatley, I., and Werner, M., *Astrophys.J.*, **258**, 135-142 (1982).
26. Henry, J. P., Depoy, D. L., and Becklin, E. E., *Astrophys.J.Lett.*, **285**, L27-L30 (1984).
27. Rieke, G. H., Lebofsky, M. J., *AIP Conf. Proc.*, **83**, 194-203 (1982).
28. Storey, J., Allen, D. A., *M.N.R.A.S.*, **204**, 1153 (1983).
29. Brown, R. L., Johnston, K. J., and Lo, K. Y., *Astrophys.J.*, **150**, 155 (1978).
30. R. D. Ekers, presented at the Galactic Center Workshop at Caltech, Jan. 1984.
31. Forrest, W., *et al*, this volume.

32. Allen, D. A., this volume.
33. Frank, J., *M.N.R.A.S.*, **187**, 833 (1979).
34. Allen, D. A., Sanders, R. H., *Nature*, **319**, 191 (1986).
35. Rieke, G. H., this volume.
36. Sellgren, K. *et al*, this volume.
37. McGinn, M. *et al*, this volume.
38. Sargent, W. L. W., this volume.
39. Gurzadyan, V. G., Ozernoi, L. M., *Astron. & Astrophys.*, **86**, 315 (1980).
40. Hall, D. N. B., Kleinman, S. G., and Scoville, N. Z., *Astrophys.J.Lett.*, **262**, L53-L58 (1982).
41. Ozernoi, L. M., this volume.
42. Ramaty, R., and Lingenfelter, R. E., *Highlights of Astronomy*, **6**, 525-529 (1983).
43. Blandford, R. D., *AIP Conf. Proc.*, **83**, 177-179 (1982).
44. Lacy, J. H., *et al*, this volume.
45. Lo, K. Y., Claussen, M. J., in preparation.

THE IONIZED GAS IN THE GALACTIC CENTER

T. R. Geballe
United Kingdom Infrared Telescope, Hilo, HI 96720

ABSTRACT

This paper reviews recent observations and interpretations of the structure and kinematics of the ionized gas within 2 parsecs of the galactic center. The ionized gas exists largely within several arm-like features, each of which contains a great deal of internal structure. Two of the arms lie on the inside edge of the rotating ring of neutral material at a radius of 2 pc and take part in that rotation. Two other features, the northern arm and the bar, pass much closer to the center. Their kinematics are not as well understood, although the northern arm appears to define a trajectory of infall towards the IRS 16/Sgr A* complex. A small, but spatially extended region of high velocity gas is found at the location of the above complex. This gas is probably undergoing radial motion and apparently is distant (~0.05 pc) from the source of its motion; it may be the interface between gas in the bar and an as yet undetected outflowing wind from an active source in the nucleus.

INTRODUCTION

This article does not purport to be a thorough review of the ionized gas in the galactic center. Several recent and comprehensive reviews are available[1,2,3]. Instead the paper is concerned mainly with the most recent observational developments and their effect on our understanding of the galactic center.

It is not necessary to emphasize the importance of the ionized gas in helping us solve the ultimate mystery of the galactic center: the nature of the object or objects which provide the bulk of its luminosity, and possibly its mass as well. The ionized gas provides the most direct link to the highly obscured ultraviolet radiation from these objects, which bathes the central few parsecs. Only ionized gas can exist in the immediate vicinity of these objects. The UV sources may of course reveal their locations via their near-infrared continua, but it is by observing the radiation from the nearby ionized gas, and in particular its line radiation, that we can best gauge their levels of activity and the level of activity around them.

In recent years there have been steady improvements in observational techniques used to study the ionized gas; these have occurred largely in the areas of large beam far-infrared spectroscopy and in small beam mapping and spectroscopy at infrared and radio wavelengths. In particular, high resolution VLA maps[3,4] not only have provided valuable new data, but also have led to a change in the way many of us perceive the ionized gas in the galactic center. Largely as a result of the VLA images we now see the ionized gas as a whole rather than as a set of randomly located ionized gas clumps. A consequence of that has been that the approach to infrared velocity-resolved spectroscopy of the ionized gas has changed from focussing on the motions of clumps to studying the motions of the patterns of ionized gas revealed by the radio images[5]. In fact, there has been an interplay between the radio continuum and infrared NeII line data, because the radio images themselves are confusing without the addition of velocity information. A second significant development has been on a smaller scale - the exploration of what one might call (in analogy to active galactic nuclei) the broad line region (BLR), located near the position of the compact radio object and IRS 16[6,7,8].

In spite of these improvements, a large number of highly divergent suggestions have been made recently regarding the nature of the active source (or sources) and the type of kinematic activity of the ionized gas. Many of the most recent of these are summarized in Table I. The proliferation of these interpretations can only be ascribed to the inventiveness of the human mind when confronted by large quantities of detailed, but somewhat ambiguous data. It is clear that further and more discriminatory measurements are required. In the infrared region some of the techniques for making them are nearly in hand. Indeed, this conference marks the transition from infrared observations of the galactic center made largely by single detectors to infrared data obtained primarily by arrays.

Table I

Recently Suggested Models for the
Ionized Gas in the Galactic Center

Active (UV) Source(s)	Ionized Gas Motion	Ref.
OB stars	-	9
"exotic object"	precessing jets	10
OB stars	infall (all arms)	4
"central engine"	outflow (BLR only)	11
-	" " "	8
"underfed b.h." (5×10^6 $M_\odot$)	-	12
"exotic star" or 100$M_\odot$ b.h.	outflow (BLR only)	13
"exotic star" + 100$M_\odot$ b.h.	" " "	14
4×10^6 $M_\odot$ b.h.	rotation and infall	15
" "	" " "	5
-	infall with drag: N arm	16
B stars + b.h. <100$M_\odot$	-	17
>10^5 $M_\odot$ object	expanding magnetic loops	18
-	expanding bar and N arm	19

STRUCTURE OF SGR A WEST

In large scale radio maps of the galactic center region the morphology of Sgr A appears considerably different from that of its neighboring radio continuum regions, Sgr B and C. The latter two regions contain many compact and bright clumps of emission. These are signposts of massive star formation - the clumps of ionized gas contain hot stars which are ionizing their natal clouds. Such compact, high contrast clumps are not seen towards Sgr A. At long (~20 cm) wavelengths this is partly because the emission from Sgr A West, the thermal H II region at the galactic center, is overwhelmed by radiation from Sgr A East, a supernova remnant which lies close to the galactic center and along the line of sight. However, at shorter wavelengths and at high angular resolutions, where the contribution from Sgr A East is insignificant, the unique structure of Sgr A West is clearly revealed. As shown in Fig. 1, the ionized gas is arranged in elongated, coherent structures, some of them quite filamentary in nature[3], with very little gas elsewhere. Far

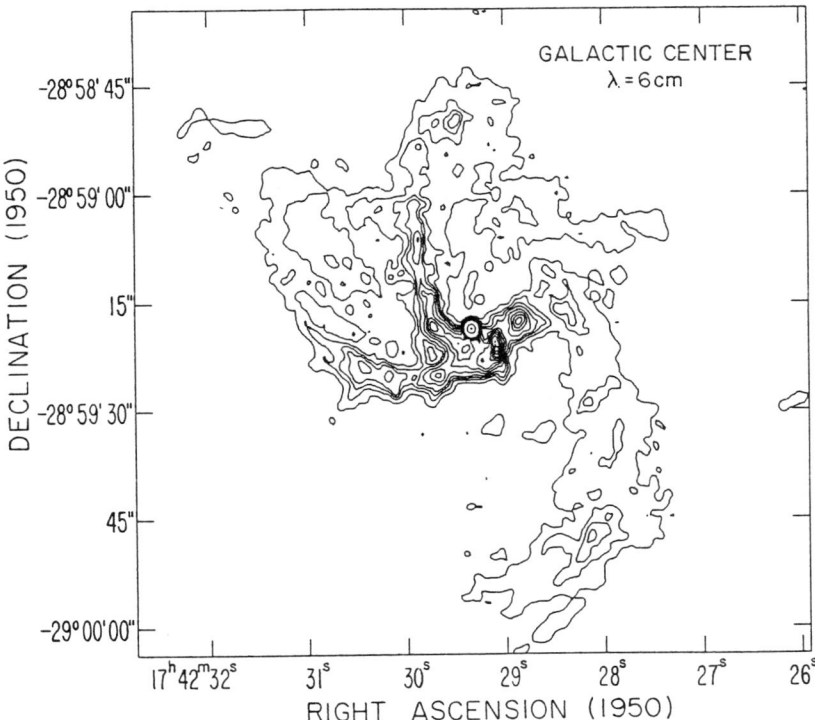

Figure 1. VLA 6cm map at 1" resolution of Sgr A West[4] (1pc ≈ 25"). The compact nonthermal radio source (Sgr A*) is situated at the center of the map on the northern edge of the bar.

infrared measurements of forbidden lines also point out this dichotomy of ionized gas densities[20,21]. It is noteworthy that the IRS16/SgrA* complex, although located roughly at the center of mass of the ionized gas, is not deeply embedded within it.

One would not expect to see a star formation morphology in Sgr A West, for the simple reason that within the dense and rapidly rotating nuclear stellar cluster interstellar gas cannot be bound to any one star for long. It has been suggested that a cluster of hot and massive stars did form in the galactic center during a previous epoch and that the remnants of this cluster are the present ionizing sources in Sgr A West[22]. Certainly it is difficult to envision star formation occurring within the turbulent environment which we see today. If the ionizing radiation for Sgr A West is supplied by hot stars, those stars must have formed roughly 10^7 years ago and under vastly different conditions than we see today. A cluster age of 10^7 years appears to be required in order to account for the both the large amount of ionized gas in Sgr A West and its low ionization state[22], which, for a standard initial mass function, imply a higher ionization state than is observed, unless the hottest and most massive stars in the cluster have evolved to or past the red giant stage. In such a scheme one might expect to find the spatial distribution of the more massive red (super)giants to be more centrally condensed than that of the less massive objects which are the current sources of ionizing photons. This may not be the case. Many or all of the UV sources are thought to lie close to, or within, the IRS16/SgrA* complex[23], which is believed to be approximately coincident with the center of the nuclear stellar cluster[24,25]. The nearest red (super)giant to IRS16 is at least 0.25 pc (6") distant.

DYNAMICS OF THE ARMS

Although, the radio images demonstrate that on a coarse scale the ionized gas exists in arm-like features, the overlaps and near overlaps of the arms make it difficult to determine where one arm ends and another begins. Information on the velocity distribution of the gas is required to determine the actual morphology and has been provided largely by measurements of Ne II 12.8μm line profiles[5,26]. At present the major structural features of the ionized gas are fairly well delineated. The relatively faint western arm stretches from the bottom to the top of Fig. 1 and is, in fact, the inside surface of a tilted, 2 pc radius ring of dust and molecular gas[27,28]. The ionized gas on the ring surface is in nearly circular orbit about the galactic center; its motion implies an enclosed mass of $\sim 5 \times 10^6$ $M_\odot$. Other distinct and brighter features that are seen in the 6cm radio map are the northern arm, running in a southerly direction from the top of Fig. 1 towards the center, and the bar, which runs roughly east-west in the center of the figure. The eastern inner edge of the ring is not apparent on the VLA 6 cm map in Fig. 1, but it is visible on a 1.3 cm map[29] made at Nobeyama (Fig. 2). The relative lack of ionized gas on the eastern side of the ring parallels the distribution of molecular line emission, suggesting a lack of ring material there. It also is possible that the eastern edge is better shielded from ionizing photons; this is suggested by the position of IRS 16/Sgr A* with respect to the nearby ionized gas in the northern arm and bar (see Fig. 1).

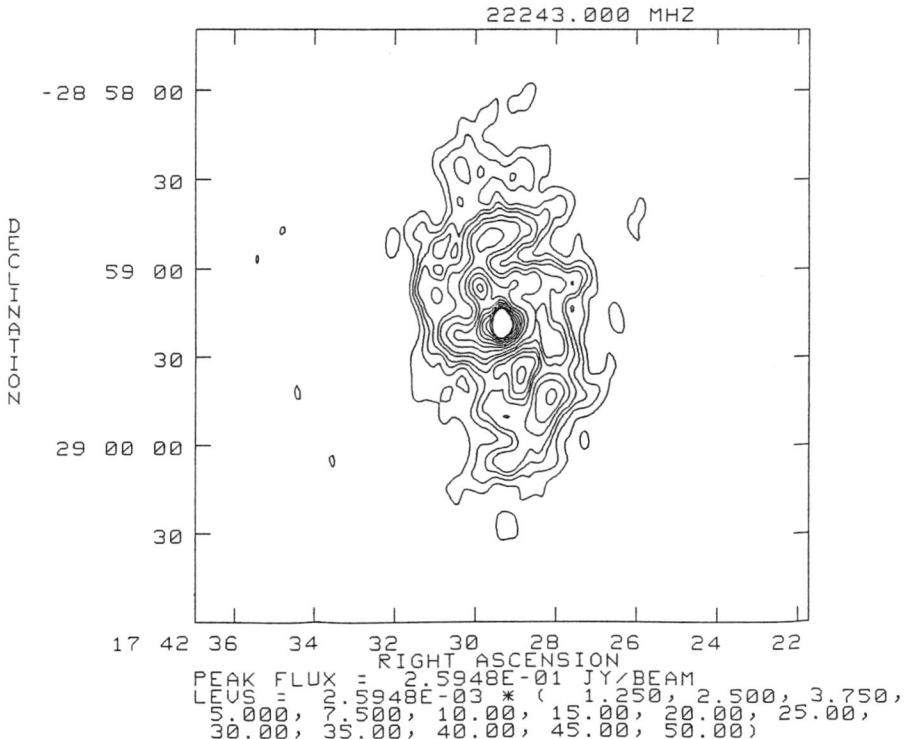

Figure 2. Map of the galactic center at 1.3 cm, made at Nobeyama[29]. The angular resolution is 6".

Like the western arm, the northern arm appears to define a trajectory; however, in this[5] case the trajectory generally is believed to be one of infall[5]. Indeed, the lack of gas on the eastern side of the ring seems to suggest that previously the northern arm was attached[5] to that part of the ring. A free-fall model of the northern arm[5] leads to the conclusion that the mass distribution of the inner two parsecs of the Galaxy includes a central compact object of mass $3-4 \times 10^6$ $M_\odot$ located close to or at IRS 16, (see Fig. 3). Successful free-fall models in which only a stellar cluster is present require considerably larger masses, which violate constraints imposed by the orbiting ionized gas in the western arm. However, once drag from a dilute inter-arm medium is introduced, the distinction between cluster and compact massive source becomes much hazier[16]. The proper interpretation of the motion of the northern arm is uncertain at present.

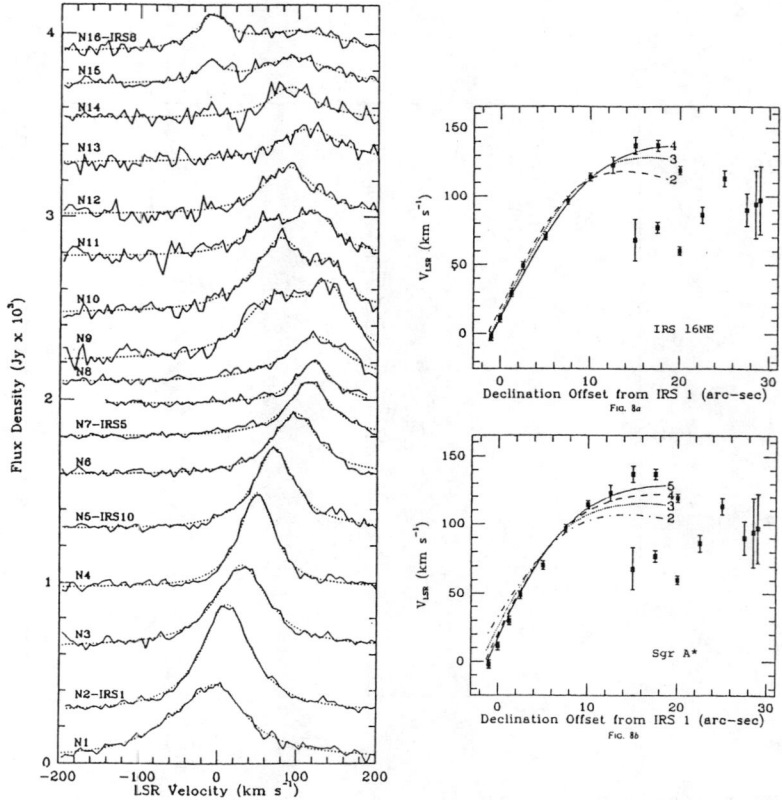

Figure 3. Spectra of the Ne II line along the northern arm and their fit to a trajectory of free infall[5]. Numbers in right hand panels refer to masses (in units of 10^6 $M_\odot$) of compact object located at IRS 16NE and Sgr A*.

The bar of ionized gas, which runs roughly perpendicular to galactic equator for about 1 parsec and passes a few arc-seconds to the south of IRS16 and Sgr A*, contains the brightest radio continuum emission from the ionized gas, suggesting proximity of the bar to the dominant source(s) of UV radiation. The ionized gas also has an unusually high kinetic temperature, which is much higher than that of the gas in the arms[1,30]. The range of observed velocities in the bar is greater than anywhere else in Sgr A West except in the broad line region. The Ne II line profiles are complex[26], which makes it difficult to follow gas motions across the bar. Recent velocity-declination plots of the bar and its surroundings[31] suggest that one arm of gas in the bar, running from IRS 1 to IRS 2, connects with the northern arm at IRS 1. However, if these two features form one coherent structure, non-gravitational forces must be invoked in order to explain the sudden changes in direction and velocity gradient at IRS1. Other velocity features of the ionized gas in the bar remain to be interpreted.

THE BROAD LINE REGION

Early velocity-resolved spectroscopy of the ionized gas near the IRS 16/Sgr A* region failed to reveal anything unusual. Then in 1982 the He I singlet line at 2.06μm was found to be exceptionally broad, even for the bar[6]. Subsequently, the same high velocities were found in the Brα 4.05μm line as are present in the helium line[7]. The strength and breadth of the helium line implies that the high velocities constitute an organized motion[7,32]. As illustrated in Fig. 4, the broad wings of the Brα line are faint relative to the central "narrow" (FWHM = 250 km/s) line emission from the bar. The wings are considerably more prominent in the helium line[7]. The enhancement of the helium line wings might be accounted for by any of several mechanisms. One possibility is a transfer of helium atoms from the 2S triplet to the 2P singlet state by electron collision, a density-sensitive process[7,33]. The enhancement also could be the consequence of ionization equilibrium, the helium being more fully singly ionized in the broad line region. Finally, there is the possibility that the enhancement reflects an actual helium overabundance (He/H ~ 1/4); this would imply that the broad line region contains outflowing material from an unusual object.

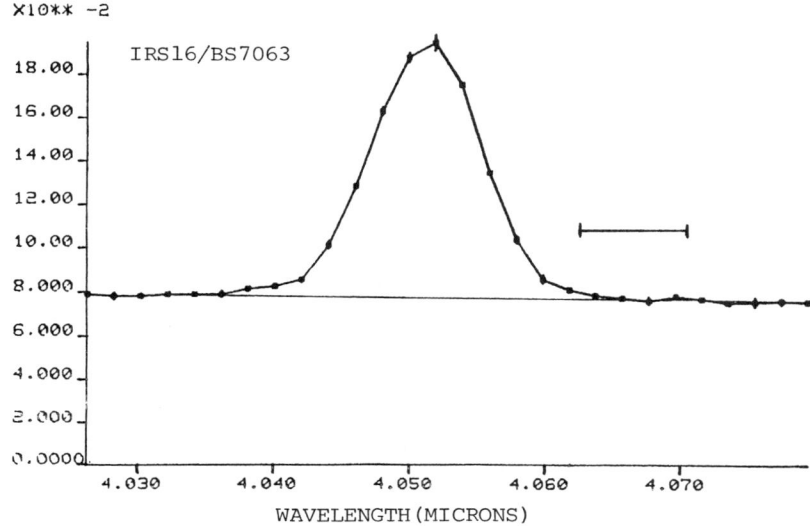

Figure 4. Spectrum of the Brα line at IRS 16, obtained at UKIRT on 10 October 1986. The velocity resolution (FWHM) of 550 km/s is indicated.

The high velocity wings are not just an extension of the underlying velocity profile of the bar, they are a separate component of the ionized gas. At UKIRT the Brα line has been mapped around IRS 16 with a 2.5" beam and with sufficient velocity resolution that the broad component could be isolated. Figure 5 shows the distribution of both the total line intensity and the intensity of the high velocity redshifted gas. Note that the two intensity distributions have little in common and that the distribution of the total line intensity bears a strong resemblance to the 6 cm VLA map near Sgr A* (Fig. 1), as it does to other Brα maps[34]. These data and others[8] demonstrate several important properties of the broad line region:
1. the high velocity gas is centered on IRS 16 (in fact on IRS 16 Center), not on the compact radio source;
2. the broad line region is spatially resolved; it has a characteristic diameter at half-maximum (deconvolved) of ~3" or ~0.15 pc.; and
3. there is no evidence for rotation or bipolarity in the high velocity gas.

Although these results suggest that the observed ionized gas in the broad line region is undergoing radial motion, that motion probably is not a free gravitational infall onto or outflow from a single unresolved object. Unless that object totally dominates the mass distribution of the central 0.1 - 0.2 pc, such broad recombination line emission will appear centrally condensed (pointlike). For a pointlike distribution not to occur, radial

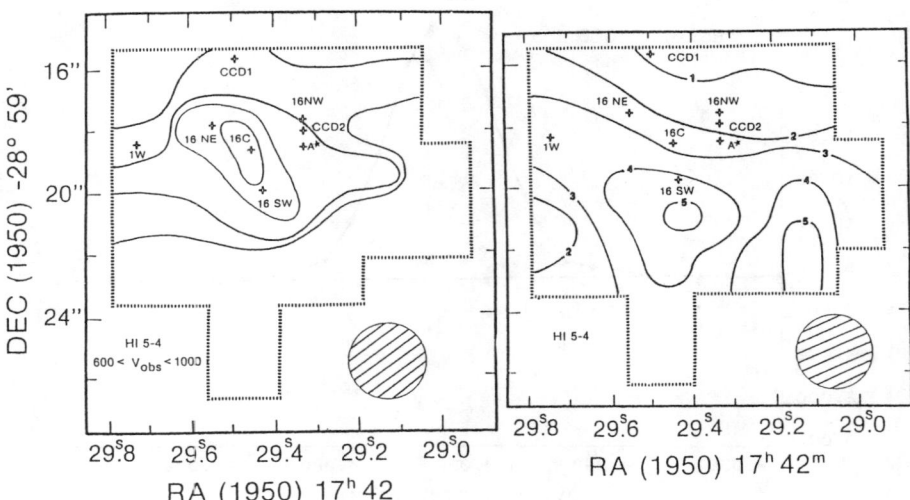

Figure 5. Maps of the Brα line flux in a 2.5" beam. Locations of various compact objects are indicated. a) Line flux from the high velocity redshifted gas; contours are every 3.1×10^{-21} W/cm^2. b) Total line flux; contours are every 1.0×10^{-19} W/cm^2.

velocities must decrease with increasing distance from the object as or more rapidly than $r^{-0.5}$ (an $r^{-0.5}$ dependence would be caused by the gravitational field of such a dominant object). Given the apparent cut-off in line emission at velocities higher than ~ 700 km/s, together with the large spatial extent of the broad line region, it is difficult to understand the observations as a flow proceeding from or to a single source, even if the source is massive. It should be noted that there are several other reasons for doubting the existence of a dominant point-mass at the galactic center[13,17,35]. It seems likely that if only a single source or sink is involved, the observed high velocity gas must be remote from that object and be a secondary effect of its activity.

Thus, the simplest models of the broad line region do not appear to work, and other models should be sought. One possibility is that the region is produced by more than one high velocity radial flow, and that the resultant pointlike high velocity Brα regions were not spatially resolved in the 2.5" beam (i.e., the broad line region only appears to be spatially extended). This interpretation

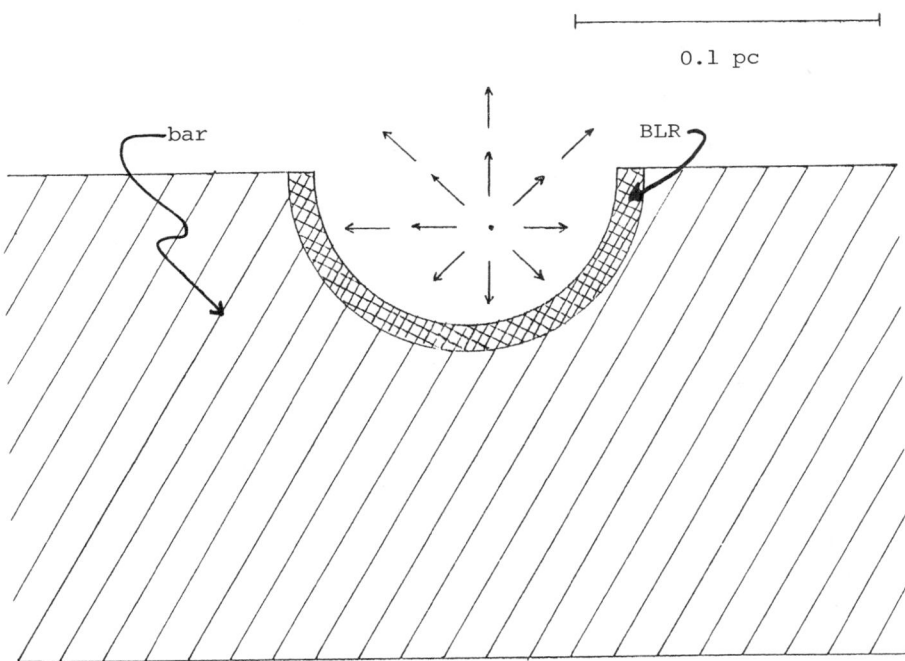

Figure 6. Possible model of the broad line region. The sketch shows a high velocity ionized outflow colliding with ionized gas in the bar (hatched). The observed broad lines are emitted in the interaction zone (crosshatched), where the wind is strongly decelerated.

requires the simultaneous existence of at least three high velocity outflow sources in the IRS 16 region, and probably more if the outermost (~1.5σ) contour in Fig. 5a is to be believed.

If the broad line-emitting region actually is extended, then its mass flow rate is ~10^{-3} $M_\odot$/yr, and the rate of momentum deposition is ~1 $M_\odot$ km/s yr. As pointed out above, most of the line-emitting gas must then be physically remote from the object which is causing the line emission. A way of producing this situation might be via a much higher velocity mass-loss wind, which is strongly decelerated as it encounters higher density gas in the bar, with the observed broad lines arising in an interaction zone between the outflowing gas and the bar (see Fig. 6). Such a wind carries momentum in larger chunks per particle than the observed gas in the broad line region, making it more difficult to detect as a pointlike source of recombination lines. Given the difficulty of detecting the observed broad lines, the present non-detection of this higher velocity wind does not appear to constrain the model. An attractive aspect of this model is the way in which it links many of the observed phenomena in the bar and the broad line region. The interaction of the high velocity wind and the gas in the bar may contribute to the high electron temperature of the bar. The location of the outflow source on the plane of the sky is not severely restricted; the source could be any of the compact infrared objects within the IRS 16 complex, and it even could be Sgr A*. In this regard it is worth noting that no connection between Sgr A* and its interstellar environment has been established. It would be surprising if the two most unusual phenomena in the galactic center, Sgr A* and the broad line region, were not related in some manner.

CONCLUSION

The ionized gas in the galactic center forms a unique HII region in our galaxy. Considerable progress has been made in observing its myriad of fascinating details and in understanding them, particularly those on a large (~1 pc) scale. Yet a convincing and unifying model of the ionized gas within ~ 0.5 pc of the center remains elusive. In large part this is because the nature of the ionizing sources (their locations, masses, luminosities, and wind parameters) are uncertain. It may be that the ionized gas in the central parsec is affected by several phenomena, none of which is dominant and, thus, that a simple model involving a single, environment-controlling object, is inappropriate.

It is informative to compare Sgr A West with the ionized gas in active extragalactic nuclei, despite the different types and scale sizes of the data. The galactic center occasionally is referred to as an example of Seyfert nucleus, albeit a very weak one. In this paper the highest velocity gas in Sgr A West has been said to define a "broad line region". If this high velocity gas is spatially extended, the broad line region has a sufficiently low density that forbidden lines should be detectable from it; thus it would not fit the definition as applied to Seyfert 1 galaxies. On the other hand, the breadths of the Brα and HeI lines in the broad line region excede the linewidths observed in classical Seyfert 2 nuclei. Nevertheless, Sgr A West, when viewed as a whole, bears a much stronger resemblance to a Seyfert 2 nucleus than to a Seyfert 1 nucleus, because its broad line emission contributes much less than one percent of the total line emission. Recently, high sensitivity optical measurements of some Seyfert 2-like galaxies have revealed that they too contain very weak emission at high velocities[36]. Both lines of evidence suggest that within every Seyfert 2-like nucleus may lurk a weak broad line region.

REFERENCES

1. Brown, R.L. and Liszt, H.S., Ann. Rev. Astr. Ap. 22, 223.
2. Genzel, R., Proceedings of the 1986 NATO Summer School "The Galaxy", eds. G. Gilmore and R. Carswell (Reidel, Dordrecht, 1986), in press.
3. Lo, K.Y., Science 233, 1394 (1986).
4. Lo, K.Y. and Claussen, M.J., Nature 306, 647 (1983).
5. Serabyn, E. and Lacy, J.H., Ap.J. 293, 445 (1985).
6. Hall, D.N.B., Kleinmann, S.G. and Scoville, N.Z., Ap. J. (Letters) 260, L53 (1982).
7. Geballe, T.R., Krisciunas, K., Lee, T.J., Gatley, I., Wade, R., Duncan, W.D., Garden, R. and Becklin, E.E., Ap.J. 284, 118 (1984).
8. Geballe, T.R., Wade, R., Krisciunas, K., Gatley, I. and Bird, M.C., Ap.J., submitted (1986).
9. Lebofsky, M.J., Rieke, G.H. and Tokunaga, A.T., Ap.J. 263, 736 (1982).
10. Brown, R.L., Ap.J. 262, 110 (1982).
11. Gatley, I., Jones, T.J., Hyland, A.R., Beattie, D.H. and Lee, T.J., M.N.R.A.S. 210, 565 (1984).
12. Sanders, R.H. and van Oosterum, W., Astr. Ap. 131, 267 (1984).
13. Ozernoy, L.M., Astron. Tsirkuliar No. 1342 (1984).
14. Ozernoy, L.M., Astron. Tsirkuliar No. 1349 (1984).
15. Crawford, M.K., Genzel, R., Harris, A.I., Jaffe, D.T., Lacy, J.H., Lugten, J.B., Serabyn, E. and Townes, C.H., Nature 315, 467 (1985).
16. Quinn, P.J. and Sussman, G.J., Ap.J. 288, 377 (1985).
17. Allen, D.A. and Sanders, R.H., Nature 319, 191 (1986).
18. Heyvaerts, J., Pudritz, R.E. and Norman, C.A., this volume (1986).
19. Schwarz, U.J., Van Gorkom, J.H. and Bregman, J.D., this volume (1986).
20. Herter, T., Houck, J.R., Shure, M., Gull, G.E. and Graf, P., Ap.J. (Letters) 287, L15.
21. Genzel, R., Watson, D., Townes, C., Lester, D., Dinerstein, H., Werner, M. and Storey, J., this volume (1986).
22. Lacy, J.H., Townes, C.H. and Hollenbach, D.J., Ap.J. 262, 120 (1982).
23. Henry, J.B., DePoy, D.L. and Becklin, E.E., Ap. J. (Letters) 285, L27 (1984).
24. Bailey, M.E., M.N.R.A.S. 190, 217 (1980).

25. Allen, D.A., Hyland, A.R. and Jones, T.J., M.N.R.A.S. 204, 1145 (1983).
26. Lacy, J.H., Townes, C.H., Geballe, T.R. and Hollenbach, D.J., Ap.J. 241, 132 (1980).
27. Gusten, R., this volume (1986).
28. Gatley, I., this volume (1986).
29. Ishiguro, M., Fomalont, E., Morita, K-I., Kasuga, T., Kanzawa, T., Iwashita, H., Kawabe, R. Kobayashi, H. and Okumura, S., Nature, submitted (1986).
30. van Gorkom, J.H., Schwarz, U.J. and Bregman, J.D., I.A.U. Symposium No. 106, The Milky Way Galaxy, eds. H. van Woerden et al. (Reidel, Dordrecht, 1985), p. 371.
31. Lacy, J.H., Lester, D.F., Arens, J.F., Peck, M.C. and Gaalema, S., this volume (1986).
32. Wollman, E.R., Smith, H.A. and Larson, H.P., Ap.J. 258, 506 (1982).
33. Osterbrock, D.E., The Astrophysics of Gaseous Nebulae (Freeman, San Francisco, 1974).
34. Forrest, W.J., Shure, M.A., Pipher, J.L. and Woodward, C.E., this volume (1986).
35. Rieke, G.H. and Lebofsky, M.J., this volume (1986).
36. Fillipenko, A.V. and Sargent, W.L.W., Ap.J. Suppl. 57, 503 (1985).

GALACTIC POSITRON ANNIHILATION RADIATION

R. Ramaty
Laboratory for High Energy Astrophysics
Goddard Space Flight Center
Greenbelt, Maryland, 20771

R. E. Lingenfelter
Center for Astrophysics and Space Sciences
University of California San Diego
La Jolla, California, 92093

ABSTRACT

Recent observations suggest that galactic line emission at 511 keV results from the superposition of contributions from a variable, compact source and an interstellar distribution of positrons resulting from the decay of radionuclei produced by thermonuclear burning in supernovae. The compact point source could have turned on as recently as 1977 and has not been seen since 1979. Photon-photon pair production in the vicinity of a relatively small black hole ($< 10^3$ $M_\odot$) could be the source of the annihilating positrons in the point source. It is not known whether this compact object lies exactly at the Galactic Center.

I. INTRODUCTION

The production of positron-electron pairs in variable, compact and high luminosity hard X-ray and gamma ray sources is now recognized to be a dominant process shaping the spectra of such objects[1]. In particular, pair production is thought to play a dominant role in the physics of active galactic nuclei[2]. However, the 511 keV line, which is the most unambiguous signature of positron annihilation, was detected only from the central regions of our Galaxy. In this paper we review the line and relevant high energy continuum observations and discuss their interpretations.

II. THE ANNIHILATION RADIATION DATA

Positron annihilation radiation from the direction of the Galactic Center was first reported in the early 1970's by Haymes et al. (see ref. 3) at Rice University following a series of observations with low resolution NaI instruments. A spectral feature at 476±24 keV was indicated by the first two of these observations. But this feature was not seen in the third observation, which suggested instead a peak at 530±11 keV. It was not until 1977 that the annihilation line energy of 511 keV was clearly identified[4] by observations with a high resolution Ge instrument flown by Leventhal et al. at the Bell and Sandia Laboratories. These observations showed that the line is very narrow (FWHM < 3.2 keV) and that the continuum below 511 keV contains a significant contribution from orthopositronium annihilation. It was in fact suggested[5] prior to this high resolution observation that the shifted peak at 476 keV could result from the convolution of the broad response function of the Rice University detector with a Galactic Center spectrum consisting of a narrow line and the accompanying orthopositronium continuum. This interpretation of the early Rice University data still seems valid. However, apart from instrumental uncertainty, no satisfactory interpretation exists for the 530 keV peak.

Subsequent observations by Riegler et al.[6] in the fall of 1979 and spring of 1980 with a Ge spectrometer on HEAO-3 also detected the 511 keV line. These observation provided, for the first time, strong evidence that the line flux varies with time, decreasing from $(1.85\pm0.21) \times 10^{-3}$ photons cm^{-2} sec^{-1} in the fall to $(0.65\pm0.27) \times 10^{-3}$ photons cm^{-2} sec^{-1} in the spring. This variability is particularly significant since it was established by observations with the same instrument. The time scale of the variation, less than 1/2 year, implies that the annihilation radiation was produced in a region of size less than 1/2 light year. The fall 1979 observations also indicate that the direction of this source is within $\pm4°$ of the Galactic Center. As we have argued before[7,8], the positrons responsible for the observed variable annihilation radiation were most likely produced by a single object, probably a black hole.

The variable nature of the positron annihilation source was confirmed by independent Ge detector observations[9,10,11,12]. The most significant of these are the observations carried out by the Bell/Sandia group from 1977 to 1984, since they also allow the comparison of data from essentially the same instrument. These observations yielded 511 keV fluxes of $(1.22\pm0.22) \times 10^{-3}$ photons cm^{-2} sec^{-1} in November 1977 (ref. 4) and $(2.35\pm0.71) \times 10^{-3}$ photons cm^{-2} sec^{-1} in April 1979 (ref. 9), and 2σ upper limits of $< 7.5 \times 10^{-4}$ photons cm^{-2} sec^{-1} in November 1981 (ref. 10) and $< 9.5 \times 10^{-4}$ photons cm^{-2} sec^{-1} in November 1984 (ref. 12). Thus, there is clear indication that the 511 keV flux decreased after 1979 and marginal evidence for an increase in flux from 1977 to 1979.

In addition to the data already mentioned, there were two other observations prior to 1979 which detected the 511 keV line. A Ge detector observation[13] by Albernhe et al. of Saclay and Toulouse and a NaI detector observation[14] by Chupp et al. at the University of New Hampshire. The fluxes measured in these observations, as well as the fluxes and upper limits discussed above, are plotted in Figure 1 as a function of the time of observation. Also indicated are the detector opening angles (FWHM in degrees). The possible correlation of the observed fluxes with both opening angle and energy resolution was pointed out by Dunphy et al.[15].

Very recently new data on galactic positron annihilation radiation has become available from observations by Share et al.[16] with the NaI gamma ray spectrometer on SMM. This satellite, launched in 1980 and still operational, has already provided[17] a wealth of data on gamma ray emission from solar flares. Since the detector points continuously at the Sun, the Galactic Center passes

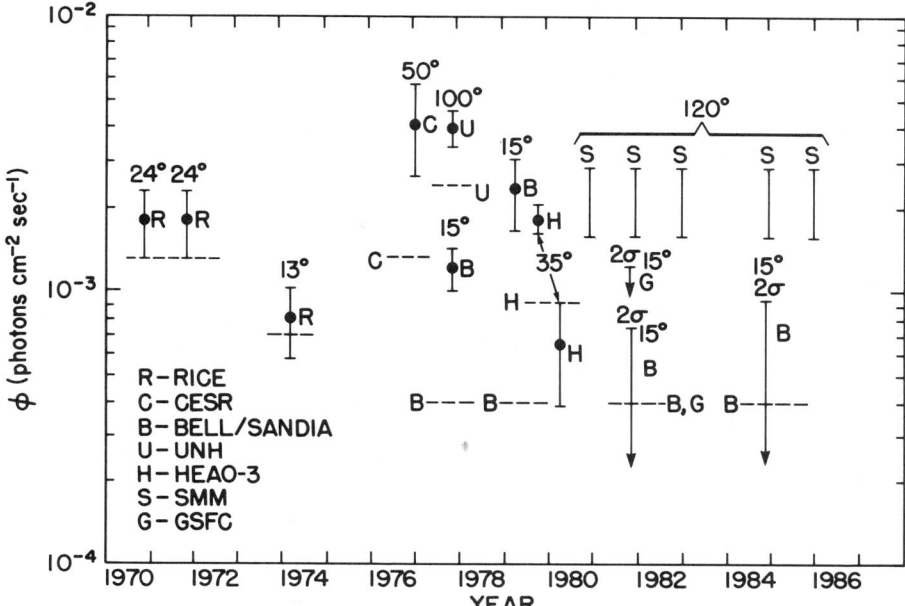

Figure 1. Observations of 511 keV line emission from the direction of the Galactic Center. The dashed lines represent the contributions of the diffuse component.

through its field of view once a year. Searching for a yearly modulation in the data, the SMM observers were able to separate a 511 keV feature of astronomical origin from the copiously produced 511 keV emission in the atmosphere, the satellite and the detector. They report[16] a flux of (1.6 to 2.9) x 10^{-3} photons cm^{-2} sec^{-1} for emission from a point source. This flux is also plotted in Figure 1 at times corresponding to the passage of the Galactic Center through the 120° field of view of the detector. There is no data for the December 1983 passage because the instrument was not fully operational just prior to the repair of SMM in space in early 1984. For the other 5 passages the line flux has not varied by more than 30% from year to year.

Unless the annihilation source varied on time scales of days, the disparity between the SMM observations and the upper limits obtained in 1981 and 1984 (see Figure 1) suggests[16] that the bulk of the emission observed with SMM is spatially distributed. If the annihilation radiation is produced with a distribution similar to that of the galactic CO, the SMM observations yield[16] an annihilation flux centered around 511 keV of (1.4 to 2.7) x 10^{-3} photons cm^{-2} sec^{-1} per radian of galactic longitude, with a best value of ~1.8 x 10^{-3} photons cm^{-2} sec^{-1} rad^{-1} (ref. 18). The possible existence of a spatially distributed galactic 511 keV emission with similar flux was suggested[15] previously from measurements up to 1980.

We have examined the consistency of a model in which the observed 511 keV fluxes are due to a variable point source at or close to the Galactic Center and a time independent distributed galactic 511 keV emission. The dashed lines in Figure 1 represent our estimates of the contribution of the diffuse component to the various observed fluxes. We discuss these estimates in more detail after we review the expected spectrum of positron annihilation radiation.

III. THE SPECTRUM OF POSITRON ANNIHILATION RADIATION

Positrons annihilating in neutral or ionized hydrogen can do so either directly or by forming bound states of positronium prior to annihilation. It is well known (e. g. ref. 19) that 25% of positronium in the ground state is parapositronium and 75% is orthopositronium, and that essentially all the positronium is expected either to be formed in the ground state or to deexcite to the ground state before annihilating. Direct annihilation produces 2 photons per positron in a line at 511 keV. Parapositronium annihilation also produces 2 photons per positron in a line at this energy, but orthopositronium annihilation produces 3 photons per positron in a continuum below 511 keV. Orthopositronium, with a lifetime of ~10^{-7} sec, can be broken up before annihilation if the density of the ambient medium exceeds ~10^{15} cm^{-3} (ref. 20) or if the radiation density above ~6.8 eV exceeds ~10^3 erg cm^{-3} (ref. 21). Orthopositronium annihilation can also be suppressed by the presence of large amounts of dust grains[22] as well as by magnetic fields >1000 Gauss (ref. 20). In the subsequent discussion we assume that the densities and fields are sufficiently low so that all the orthopositronium annihilates.

In molecular hydrogen, the line width resulting from direct annihilation is 1.56±0.09 keV, based on laboratory measurements[23] in a low density medium. The calculated[24] line width for parapositronium annihilation is ~6.5 keV, in good agreement with the laboratory measurements[25] which yield a width of 6.4±0.1 keV. In partially ionized hydrogen, the line width from both direct and parapositronium annihilation is ~2.5 keV at 5×10^4 K and increases as $T^{1/2}$ with increasing temperature[20]. All of these values are full widths at half maximum (FWHM).

The fraction f of positrons annihilating from bound states of positronium was measured[25] in low density molecular H yielding f = 0.897±0.003. In partially ionized hydrogen f exceeds ~0.9 if 10^4 K $\le T \le 5 \times 10^4$ K and the ratio of hydrogen atoms to free electrons is greater than 0.1 (ref. 24). At higher temperatures and higher ionization fractions f is substantially lower[24]. For example, in a fully ionized hydrogen plasma at 10^6 K, f $\simeq$ 0.5.

The continuum spectrum $P_t(E)$ resulting from orthopositronium (triplet) annihilation in its rest frame is given by an expression derived by Ore and Powell (see ref. 5). In Table 1 we list the values of this function at selected photon energies, E, normalized to unit integral from 0 to 511 keV. ($P_t = 0$ for E > 511 keV.)

TABLE 1   Spectrum from orthopositronium annihilation

E(keV)	$10^3 P_t(E)$ (keV^{-1})	E(keV)	$10^3 P_t(E)$ (keV^{-1})
510	4.42	420	3.13
500	4.03	400	3.00
490	3.81	350	2.67
480	3.65	300	2.32
470	3.53	250	1.94
460	3.43	200	1.55
450	3.35	150	1.15
440	3.27	100	0.76

In a warm or hot medium, P_t will be modified by Doppler shifts, but for temperatures $< 5 \times 10^4$ K the modification is not very important.

We can now write the spectrum of the annihilation radiation per annihilating positron as follows:

$$P(E) = 2 (1-f) P_{da}(E) + 0.5 f P_s(E) + 2.25 f P_t(E), \qquad (1)$$

where $P_{da}(E)$ and $P_s(E)$ are the line profiles resulting from direct annihilation and parapositronium (singlet) annihilation, respectively. Both these function are normalized to unit integral. These two functions can be approximated by Gaussians of widths as given above. For annihilation in molecular hydrogen, the total line spectrum (resulting from the first two terms in equation 1) can be approximated[25] by a single Gaussian with width ~2.2 keV. It was pointed out[25] that this width is in good agreement width the width of 1.6(+0.9-1.6) keV observed[6] in the fall of 1979. This observed width is also in good agreement with the width expected from annihilation in partially ionized hydrogen if $T < 5 \times 10^4$ K.

Orthopositronium annihilation continuum was seen in the 1977 Bell/Sandia data (ref. 4) with $f \simeq 0.9$, in the 1980 HEAO-3 data (ref. 26) with $f \simeq 1$ and in the 1977 UNH data with $f \simeq 0.7$ (ref. 14). In all of these observations, a sharp cutoff of the continuum was observed at energies >511 keV. The shifted peak at ~476 keV observed[3] in 1970 and 1971 was explained[5] using $f = 1$. A positronium fraction $f \simeq 0.7$ was inferred[26] from the 1979 HEAO-3 data, but in this case a strong continuum was observed at energies >511 keV and therefore the deduced f depends crucially on the extrapolation of this continuum to energies <511 keV. For the continuum spectrum calculated by McKinley[27], the value of f for the 1979 HEAO-3 data is consistent with 0.

We have estimated the contribution of the diffuse 511 keV source to the various observations shown in Figure 1 as follows. We assume that the diffuse source consists of a narrow line (FWHM $\lesssim$ 2.5

keV) and an orthopositronium continuum with f = 0.9. The diffuse flux of 1.8×10^{-3} photons cm^{-2} sec^{-1} rad^{-1} observed with the SMM NaI detector (energy resolution ~35 keV at 511 keV) could therefore contain a significant contribution from orthopositronium continuum. We estimate that the narrow line flux is ~ 1.5×10^{-3} photons cm^{-2} sec^{-1} rad^{-1}. For the Ge detector observations (Bell/Sandia, HEAO-3, CESR and GSFC) we simply scale this narrow line flux with the opening angles of the detectors as indicated in the Figure 1. For the UNH NaI detector, which has an opening angle of 100°, we scale the total SMM NaI detector flux of 1.8×10^{-3} photons cm^{-2} sec^{-1} rad^{-1} with an angle of 80°, corresponding to the effective longitudinal extent of galactic high energy gamma ray emission, which tracks the CO emission. For the Rice University data, we assume an energy resolution of 116 keV, which is the value needed[5] to shift the peak of the count rate spectrum to 476 keV. Then we calculate the total diffuse emission in a band extending from 418 keV to 511 keV. We also scale the flux with the opening angles of the Rice University detectors shown in Figure 1.

The results are shown by the dashed lines in Figure 1. We see that the 1977 and 1979 Bell/Sandia observations, the 1979 HEAO-3 observation, and the UNH and CESR observations, are consistent with a point source contribution in excess of the diffuse emission. On the other hand, all the observations since 1980, including the 1980 HEAO-3 observation, as well as the observations prior to 1977, are consistent with emission from the diffuse source only. Thus, the variable point source, having an average narrow line flux of ~10^{-3} photons cm^{-2} sec^{-1}, could have been a transient lasting for approximately 2 years, from 1977 to 1979. We emphasize, however, that the decisive evidence for the point source comes not from the finite flux above our estimated diffuse component contribution but from the observed variability, as discussed in the previous section.

IV. THE CONTINUUM HARD X-RAY AND GAMMA RAY DATA

The X-ray and gamma-ray continuum observations of the Galactic Center region up to 1982 were reviewed[28] in detail. Here we highlight those observations which seem particularly relevant to the origin of the 0.511 MeV annihilation line, and discuss new data that have become available since then.

In the energy range around a few keV, Einstein observatory measurements[29] with angular resolution of ~1' revealed several sources in a region of angular size ~1° around the Galactic Center. The luminosities of these sources in the few keV range are generally smaller than a few times 10^{35} erg sec^{-1}.

In the energy range from about 10 to 100 keV, the Galactic Center region was observed with a variety of detectors, including the HEAO-1 low energy detector (LED)[28] which had an angular resolution of 1.6°. Observations with this instrument showed that the source region GCX, located within ~0.5° of the Galactic Center (ref. 30), was highly variable, particularly at the higher energies. At ~100 keV the flux was observed to vary by more than a factor of 3 in 6 months, while at ~20 keV the variability in 6 months was less than 20%. In its highest observed state GCX accounted for ~50% of the total flux at 100 keV from a region of ~10° angular size around the Galactic Center. The maximum luminosity of GCX in the 10 to 100 keV range was ~5 x 10^{37} erg sec^{-1}. Here and in the following we use 8.5 kpc for the distance[31] to the Galactic Center.

In the energy range from 100 to 500 keV angular resolutions of ~15° were achieved in a number of observations (see ref. 28). In particular, observations with the HEAO-1 medium energy detector (MED), which had 16° angular resolution, showed that at ~300 keV the flux from the direction of the Galactic Center decreased by about a factor of 4 in 6 months. The luminosity of the Galactic Center region in this energy range was ~7 x 10^{37} erg sec^{-1}.

The continuum observations which are the most relevant to the origin of the 511 keV annihilation line are the HEAO-3 observations[26]. The same Ge detectors which observed the 511 keV line in the fall of 1979 and spring of 1980, also observed the continuum from ~50 keV to ~3 MeV. The angular resolution of this detector was ~35°. While the line flux decreased by about a factor of 3 in six months from the fall to the spring, the continuum at energies > 511 keV decreased by more than a factor of 7, from 10^{-2} photons cm^{-2} sec^{-1} in the fall, to <1.4 x 10^{-3} photons cm^{-2} sec^{-1} in the spring. (The latter value is a 2σ upper limit). On the other hand, at energies < 511 keV the continuum observed by the HEAO-3 instrument was much less variable, decreasing in 6 months by only ~30% at 300 keV and ~15% at 100 keV. These variations were smaller than those observed with the HEAO-1 detectors at the same energies. Since the HEAO-1 LED and MED instruments had better angular resolutions (1.5° and 16°, respectively) than the HEAO-3 instrument (35°), the HEAO-3 observations at energies <511 keV must have included contributions from multiple sources or a spatially distributed source. The very large variation observed with HEAO-3 at energies >511 keV, however, strongly suggests that the continuum at these energies is produced by a single source.

This point source of continuum emission is probably also the source of the positrons responsible the observed 511 keV line. The direction of this source is not known, except that it is located within the 35° opening cone of the HEAO-3 instrument. On the basis of the similarity of the variability of the > 511 keV source with the variabilities observed at lower energies, it is reasonable to assume that this direction is within ~1.5° of the Galactic Center. The luminosity of the > 511 keV continuum source (in the 0.5 to 3 MeV range) was ~2 x 10^{38} erg sec^{-1}.

The Galactic Center region was also observed in the 100 MeV to 1 GeV range (see ref. 28). No variability was seen in this energy regime.

V. THE NATURE OF THE GALACTIC CENTER POINT SOURCE 511 KEV EMISSION

We shall consider first the nature of the point source of 511 keV annihilation radiation from the direction of the Galactic Center and then turn to that of the diffuse galactic component. Using f=0.9, the observed flux from the point source, ~10^{-3} photons cm^{-2} sec^{-1}, implies an annihilation rate of ~1.3×10^{43} positrons sec^{-1} or an annhilation radiation luminosity of ~2×10^{37} ergs sec^{-1}.

The nature of the positron annihilation region is constrained by the observed line width and intensity variations. The line width (FWHM < 2.5 keV) requires a gas temperature in the annihilation region less than 5×10^4 K and the flux variations require that the density of gas be >10^5 H cm^{-3} so that the positrons can slow down and annihilate in less than half a year. The direct correlation between variations in the annihilation line flux and the continuum flux at higher energies argues strongly against the possibility[22,32] that the variation in the point source line intensity was caused solely by a change in the nature of the annihilation region, affecting the rate of annihilation, and not by a change in the positron production rate at the source.

The nature of the positron source is also strongly constrained by the observed variation of the 511 keV intensity and by observations at other wavelengths. The decrease of a factor of three in the line intensity in six months clearly excludes any of the multiple, extended sources, such as cosmic rays, pulsars[33], supernovae[34] or primordial black holes[35], previously proposed. Instead, it essentially requires a single, compact (<10^{18} cm) source which is inherently variable on time scales of six months or less.

The correlation between the line and higher energy continuum intensities also argues strongly against the possibility[32] that the point source positrons result from the decay of radionuclei produced by thermonuclear burning in a single massive supernova explosion, since the higher energy continuum emission should not be correlated with the line emission in such a case.

We have previously reviewed[8,21] the various possible positron production processes and the observational constraints on them. We found that the observations[28] of the accompanying continuum emission at energies >511 keV appear to set the strongest constraints on the positron production process, requiring a very high efficiency such that about 10% of the total radiated energy >511 keV goes into electron-positron annihilation radiation. Under the conditions that the positron production occurs on time scales comparable to that of the observed variation and in an isotropically emitting region, only photon-photon pair production among ~MeV photons can provide the required high efficiency. Moreover, the absolute luminosity of the annihilation line requires that the photon-photon collisions take place in a very compact source (d < 5×10^8 cm). Pair production in an intense radiation field around an accreting black hole of <10^3 $M_\odot$ appears to be a possible source.

If the emission is not isotropic, the positrons could also result from pair production by small angle photon interactions in a beamed electromagnetic cascade generated by dynamo action in a magnetic field accreting onto a much more massive (~10^6 $M_\odot$) rotating black hole. But in such a case one should expect to see some variation in the much higher energy gamma ray emission and this has not been reported.

VI. THE NATURE OF THE DIFFUSE GALACTIC 511 KEV EMISSION

As we discussed above, recent analysis of the SMM and other observations suggest that there is a diffuse galactic component of the 511 keV line emission with an intensity of around 1.5×10^{-3} photons cm^{-2} sec^{-1} rad^{-1} of galactic longitude in the direction of the galactic center. This emission can account for all of the annihilation radiation observed from the direction of the Galactic Center after 1979 and much of that before that time. The high resolution HEAO-3 observations in the spring of 1980, which appear to be entirely of diffuse origin, also show a positronium continuum component, implying that most of the diffuse positrons may be annihilating via positronium. Since diffuse line emission has not yet been detected from the direction of the galactic anticenter, the flux in that direction is probably at least a factor of three lower. Thus, the intensity of the line emission must vary significantly with galactic longitude.

The measurement of the diffuse annihilation radiation is of great importance to our understanding of galactic nucleosynthesis. For although there are a variety of sources of positrons in the galactic disk, we have shown[7] previously that the dominant source should be positrons from the decay to ^{56}Fe of ^{56}Co, resulting from explosive nucleosynthesis in supernovae. Type I supernovae should be the major source of such positrons, if models[36-38] of their optical emission are correct. These models suggest that the energy for the optical emission comes from ^{56}Ni and ^{56}Co decay, which heats the gas in the expanding nebula. These models also suggest that the deviation of the observed light curves from the decay curve of ^{56}Co results from partial escape of positrons from the nebula, carrying away part of the decay energy. Because of the much greater mass and slower expansion velocity of the nebulae of Type II supernovae, the ^{56}Co decay positrons produced there should nearly all annihilate before the nebulae become transparent with only a negligible fraction escaping.

Once they escape into the interstellar medium, the positrons should have a mean life against annihilation of about 10^5 yrs (see ref. 24), so that positrons from over a thousand supernovae should combine to produce the steady observed disk component of annihilation radiation. If $M/M_\odot$ solar masses of ^{56}Ni are produced per supernova, and a fraction, ϵ, of the positrons from the resultant ^{56}Co decay escape from the supernova shell, for f=0.9 the expected diffuse galactic 511 keV line emission from the inner part of the Galaxy should be $\sim 0.2(M/M_\odot)$ ϵ photons cm^{-2} sec^{-1} rad^{-1} for a Type I supernova rate of 0.02 yr^{-1}. Taking $M/M_\odot$ to be ~ 0.5 from optical[39] and infrared[40] observations, the observed diffuse 511 keV flux of around 1.5×10^{-3} photons cm^{-2} sec^{-1} rad^{-1} implies a positron escape fraction ϵ of ~ 1.5%. This should be compared with the previously estimated[36] value of 10%.

Although diffuse galactic gamma ray line emission at 1.809 MeV has also been observed[41,42] from decay of the long lived positron emitting radioisotope ^{26}Al, the positrons from such decay can not account for the diffuse annihilation radiation. The ratio of positron to 1.809 MeV photon emission is 0.85 in the ^{26}Al decays, and for $f \approx 0.9$ about 0.65 line photons at 511 keV are produced per positron. Thus, the observed 1.809 MeV flux of $(4.8 \pm 1.0) \times 10^{-4}$ photons cm^{-2} sec^{-1} rad^{-1} implies that only ~18% of the observed diffuse 511 keV emission could come from ^{26}Al decay.

CONCLUSIONS

1. In the Galactic Center or its vincinity there is a variable point source of 511 keV line emission and > 511 keV continuum. Evidence for this object comes from flux variations deduced from the comparison of observations with the same detectors at different times. This point source has not been seen since 1979 and it may have turned on as recently as 1977.
2. There is diffuse 511 keV line emission from the galactic disk unrelated to the Galactic Center. Evidence for this emission comes from the comparison of observations with gamma ray spectrometers of narrow and broad fields of view. This diffuse source has shown no temporal variations from 1981 through 1985.
3. The location of the point source is not known precisely but indirect arguments would place it within ~1° of the Galactic Center. The correlation of the 511 keV line and > 511 keV continuum fluxes strongly suggests that the observed line flux variability is due to a variable positron source and not to variations in the annihilation site. The high ratio of the line-to-continuum luminosities implies a source size of a few times 10^8 cm for an isotropically emitting object suggesting a black hole of mass < 10^3 M_0. Such an object need not reside at the Galactic Center.
4. The diffuse line emission results from the annihilation of a galactic distribution of positrons. These positrons could be produced in supernovae, primarily from the decay of ^{56}Co to ^{56}Fe, and subsequently escape to the interstellar medium.

Acknowledgements: We wish to acknowledge G. H. Share for discussions of the recent SMM data and NASA Grant NSG 7541 for financial support.

REFERENCES

1. M. L. Burns, A. K. Harding and R. Ramaty, Positron-Electron Pairs in Astrophysics, (AIP, New York, 1983).
2. M. J. Rees, Ann. Rev. Astron. Astrophys., 22, 471 (1984).
3. R. C. Haymes et al., Ap. J., 201, 593 (1975).
4. M. Leventhal, C. J. MacCallum and P. D. Stang, Ap. J., 225, L11 (1978).
5. M. Leventhal, Ap. J., 183, L147 (1973).
6. G. R. Riegler, J. C. Ling, W. A. Mahoney, Wm. A. Wheaton, J. B. Willett and A. S. Jacobson, Ap. J., 248, L13 (1981).
7. R. Ramaty and R. E. Lingenfelter, Phil. Trans R. Soc. London A 301, 671 (1981).
8. R. E. Lingenfelter and R. Ramaty, The Galactic Center, (AIP, New York, 1982), p.148.
9. M. Leventhal et al., Ap. J., 240, 338 (1980).

10. M. Leventhal, C. J. MacCallum, A. F. Huters and P. D. Stang, Ap. J., 260, L1 (1982).
11. W. S. Paciesas et al., Ap. J., 260, L7 (1982).
12. M. Leventhal, C. J. MacCallum, A. F. Huters and P. D. Stang, Ap. J., 302, 459 (1986).
13. F. Albernhe et al., Astron. Astrophys. 94, 214 (1981).
14. B. M. Gardner, D. J. Forrest, P. P. Dunphy and E. L. Chupp, The Galactic Center, (AIP, New York, 1982), p. 144.
15. P. P. Dunphy, E. L. Chupp and D. J. Forrest, Positron-Electron Pairs in Astrophysics, (AIP, New York, 1983), p. 237.
16. G. H. Share et al., Advances in Space Research, in press (1987).
17. E. L. Chupp, Ann. Rev. Astron. Astrophys., 22, 359.
18. G. H. Share et al., Bull. Amer. Astron. Soc. 18, 978, (1986).
19. F. W. Stecker, Astron. and Space Sci., 3, 579 (1969).
20. C. J. Crannell, G. Joyce, R. Ramaty and C. Werntz, Ap. J., 210, 582.
21. R. E. Lingenfelter and R. Ramaty, Positron-Electron Pairs in Astrophysics, (AIP, New York, 1983), p. 267.
22. W. H. Zurek, Ap. J., 289, 603 (1985).
23. B. L. Brown and M. Leventhal, Phys. Rev. Lett., 57, 1651 (1986).
24. R. W. Bussard, R. Ramaty and R. J. Drachman, Ap. J., 228, 928 (1978).
25. B. L. Brown, M. Leventhal, A. P. Mills, Jr., and D. W. Gidley, Phys. Rev. Lett., 53, 2347 (1984).
26. G. R. Riegler, J. C. Ling, W. A. Mahoney, Wm. A. Wheaton and A. S. Jacobson, Ap. J., 294, L13, 1985.
27. J. M. McKinley, Proc. Third Internat. Workshop on Positron Gas Scattering, (World Scientific, Singapore, 1986), p. 152.
28. J. L. Matteson, The Galactic Center, (AIP, New York, 1982), p. 109.
29. M. G. Watson, R. Willingdale, J. E. Grindlay and P. Hertz, Ap. J., 250, 142 (1981).
30. E. Kellogg et al., Ap. J., 169, L99 (1971).
31. V. Trimble, Comments on Astrophysics, 11, 257 (1986).
32. W. R. Weber et al., Nature, 323, 692 (1986).
33. P. A. Sturrock and K. B. Baker, Ap. J., 234, 612 (1979).
34. R. Ramaty and R. E. Lingenfelter, Nature, 278, 127 (1979).
35. P. N. Okeke and M. J. Rees, 1980, Astron Astrophys., 81, 263.
36. S. A. Colgate, Astrophys. Space Sci. 8, 457 (1970).
37. W. D. Arnett, Ap. J., 230, L37 (1979).
38. S. A. Colgate, A. G. Petschek and G. T. Kriese, Ap. J., 237, L81 (1980).
39. T. S. Axelrod, Type I Supernovae, (Univ. of Texas, Austin, 1980), p. 80.
40. J. R. Graham et al., MNRAS, 218, 93 (1986).
41. W. A. Mahoney et al., Ap. J., 286, 578 (1984).
42. G. H. Share et al., Ap. J., 292, L61, (1985).

THE EVIDENCE FOR AND AGAINST THE EXISTENCE OF SUPERMASSIVE BLACK HOLES IN E GALAXIES

Wallace L. W. Sargent
Palomar Observatory, California Institute of Technology

ABSTRACT

We review the history of the controversies surrounding the interpretation of the velocity dispersion and light profile in the center of M87. The most recent theoretical work suggests that the highly anisotropic velocity ellipsoid, which had been proposed as an alternative to a central massive object, is unstable to the formation of a bar. The current observations are just consistent with either a central, essentially point, mass or with a massive star cluster of the kind proposed by Dressler. However, the most recent observations of the center of M31 and M32 are difficult to reconcile with anything but a central point mass.

INTRODUCTION

Following the work of Salpeter (1964)[1] and Lynden-Bell (1969)[2], it is widely believed that supermassive black holes are ultimately responsible for the phenomena found in "active galactic nuclei" and quasars. The black holes responsible for these phenomena are thought to have masses in the range 10^5–$10^9 M_\odot$. The Schwarzschild radius for black holes in this mass range is 3×10^{11}–3×10^{14} cm–distances too small to be resolved by remotely conceivable techniques in even the nearest galaxies. For example, at the distance of the Virgo cluster (15 Mpc) the Schwarzschild radius of a $10^9 M_\odot$ black hole subtends an angle of only 1.4×10^{-6} arcseconds, about a factor of 100 smaller than the resolution achievable by current earth-based Very Long Baseline Interferometry. Consequently, the evidence for the Black Hole hypothesis as an explanation of extragalactic violent events is very indirect.

For this reason, I shall focus in the brief time available on one particular aspect of the Black Hole problem–the evidence for and against their existence in the nuclei of certain nearby galaxies based on studies of stellar motion. This story reveals a dialogue between theory and observation which is supposed to be typical of the way in which science progresses, but which is so rare in astronomy. However, it also reveals that things are never simple. Earlier in this Symposium Dr. Ekers said something like–and I quote from memory–"Stars are not affected by the complications of gas dynamics–they should be very good probes of the mass distribution." I used to share this delusion; how I came to be enlightened forms the bulk of this talk.

THE SIMPLE STELLAR DYNAMICS OF ELLIPTICAL GALAXIES

The radial light distribution in E galaxies is well described by several empirical laws, for example, those of Hubble and de Vaucouleurs. It can also be explained very well by stellar dynamical models from a family introduced by King (1966)[3] in order to explain the star distributions in globular clusters.

King models have isothermal cores which are characterized by a central density ρ_0 and by a scale length (the core radius)

$$a = \frac{3\sigma_v}{(4\pi G \rho_0)^{1/2}} \qquad (1)$$

where σ_v is the observed stellar velocity dispersion (which is one component of the 3-dimensional velocity dispersion). The distribution function is isotropic, with an energy cutoff so that the models reach $\rho=0$ at finite radius r. Thus, if the total energy (kinetic plus potential) of a star of mass m^* is

$$E = \frac{1}{2}m^*v^2 - \frac{GM(r)m^*}{r} \qquad (2)$$

where the mass inside r is

$$M(r) = \int_0^r 4\pi r^2 \rho(r) dr \qquad (3)$$

then King writes the distribution function as

$$f(E) = A\exp(-\beta E) - A\exp(-\beta E_{esc}) \qquad (4)$$

and $f = 0$ for $E > E_{esc}$, the energy above which the star is imagined to escape from the cluster or galaxy. Thus, $\rho = 0$ at $r = r_t$, the tidal radius. As $E_{esc} \to \infty$ the models tend towards isothermal spheres. In the regime where the isothermal sphere is a good approximation $\rho(r) \sim r^{-2}$, exactly in conformity with the observed Hubble law.

As is well known, the relaxation times for the stars in galaxies is very long. In the center of a typical E galaxy $t_R \sim 10^{14}$ years. Therefore, the stars move in the overall gravitational potential of the whole system; the orbits are not influenced to any significant degree by encounters with other stars.

THE EFFECT OF A CENTRAL BLACK HOLE

It is of interest to investigate how King models are modified by the presence of a central black hole (or by any central "point" mass whose size is negligible as compared with the core radius). It is easy to see that the sphere of influence (cusp radius) of a point mass M_H is given by

$$\frac{3}{2}\sigma_v^2 = \frac{GM_H}{r_a} \qquad (5)$$

where σ_v is the velocity dispersion in the unperturbed isothermal core of the galaxy. Putting in appropriate scaling factors, we find

$$r_a = 70 \left(\frac{M_H}{10^9 M_\odot}\right) \left(\frac{\sigma_v}{200\text{km s}^{-1}}\right)^{-2} \text{pc.} \qquad (6)$$

At the distance of M87, the nearest active E galaxy in which we might look for signs of a black hole, $1''= 70$ pc so that we clearly require M_H to be several billion solar masses in order to detect an effect on the light profile from the ground.

In general, the effect on the light profile of a central point mass depends on the parameter $\mu = M_H/M_a$ where $M_a = 4\pi\rho_0 a^3$ is the mass inside the core radius. The problem of determining the stellar distribution function $f(E)$ and hence obtaining the mass distribution has been solved for the case where the gravitational potential of the black hole $\phi = -GM_H/r$ dominates. Solutions must satisfy the inner boundary condition that the stars are "eaten" by the black hole if they pass within the Schwarzschild radius. Several such solutions have been found; for single-mass stars by Peebles (1972)[4] and Bahcall and Wolf (1976)[5] and for a power-law mass distribution by Young (1967)[6]. In all physically reasonable cases the stars are drawn into a weak cusp centered on the black hole; the relation diverges from a King model for $r < r_a$. If the stars around the black hole are relaxed then the density distribution in the cusp varies as $\rho(r) \sim r^{-7/4}$. If the star distribution is unrelaxed and the mass of the black hole increases slowly, on a time scale longer than a typical stellar orbital period, then the orbital eccentricity is an adiabatic invariant. In this case the semi-major axis of the orbit contracts while the orbital shape stays the same. The resulting density dependence is $\rho(r) \sim r^{-3/2}$. Finally, if the stars are unrelaxed and the black hole mass is fixed, the stellar orbits are merely deflected as they pass the black hole. In this case $\rho(r) \sim r^{-1/2}$. In all cases $\sigma_v^2 \sim GM_H/r$ for $r < r_a$. However, close to the black hole the velocity distribution of the stars begins to depart from a Gaussian distribution and itself assumes the form of a cusp in velocity. The advent of the Space Telescope makes it possible to search for such an effect close to putative black holes; the Fourier Quotient method (Sargent et al. 1977[7]) which is often used to determine velocity dispersions in galaxies could be suitabliy modified to allow a non-Gaussian velocity distribution to be fitted to the data.

The inner edge of the cusp around the black hole occurs at some critical radius r_c where either stars are tidally disrupted (r_D) or the stars physically collide with one another (r_{coll}). For main-sequence stars of 1 $M_\odot$

$$r_D = 2.1 \times 10^5 \left(\frac{M_H}{10^9 M_\odot}\right)^{1/3} \text{pc}. \tag{7}$$

and

$$r_{coll} = 5 \left(\frac{M_H}{10^9 M_\odot}\right) \text{pc}. \tag{8}$$

For $M_H \geq 10^9 M_\odot$, r_D is less than the Schwarzschild radius, so that main-sequence stars are swallowed whole by much larger black holes. Also, for $M_H \sim 10^9 M_\odot$, $r_{coll} \sim 1/10\, r_a$.

APPLICATION TO OBSERVATIONS OF M87

Young et al. (1978)[8] found that the light profile of the nearby giant elliptical galaxy M87 could not be fitted to a King model inside the core radius ($\sim$ 10 arcseconds). After subtracting a central light spike (believed to be non-thermal in origin because of its blue color) Young et al. showed the observed light profile could be fitted by embedding a central black hole (or point mass) of $3 \times 10^9 M_\odot$. At the same time Sargent et al. (1978)[9] made observations of the radial variation of the radial velocity dispersion in M87. They found that σ_v rises rapidly inside the core radius of M87, reaching a value of 350 km s^{-1} at $r = 1''\!.5 = 110\delta_{15}$ pc from the center, where δ_{15} is the distance to M87 in units of 15 Mpc. (Hereafter we shall assume $\delta_{15} = 1$.) It should be noted that the resolution along the spectrograph slit (i.e., in the radial coordinate) of the Sargent et al. observations was only 5$''\!.4$; the radius $r = 1''\!.5$ is a luminosity weighted average value. The value $\sigma_v = 350$ km s^{-1} at $r = 1''\!.5$ was corrected to $\sigma_v = 400$ km s^{-1} on allowing for the fact that the line-of-sight passes through the outer regions of M87 where σ_v is lower than it is in the center.

In order to obtain a mass distribution for M87, Sargent et al. (1978) made use of the stellar "equation of hydrostatic equilibrium"

$$\frac{d}{dr}\left\{\rho(r)\sigma_v^2(r)\right\} = \frac{-GM(r)}{r^2}\rho_*(r). \tag{9}$$

Here $\rho_*(r)$ is the density of stars which give rise to observed luminosity, $\sigma_v(r)$ is the radial component of the velocity dispersion of these stars and $M(r)$ is the *total* mass inside radius r–not just the mass of the stars that produce the observed luminosity. It is important to emphasize that this equation is derived on the assumption that the stellar velocity ellipsoid is isotropic.

Now let $\rho_*(r) = (M/L)_* L_*(r)$ where $(M/L)_*$ is the mass-to-light ratio (to be determined) of the stellar constituent of M87 and $L_*(r)$ is the luminosity density (which can be derived from the measurements of Young et al. [1978]). Then the equation of hydrostatic equilibrium can be written in terms of logarithmic derivatives as:

$$M(r) = \frac{r\sigma_v^2(r)}{G}\left\{-\frac{d\ln L_*}{d\ln r} - \frac{d\ln(M/L)_*}{d\ln r} - \frac{d\ln\sigma_v^2}{d\ln r}\right\} \quad (10)$$

Studies of the outer parts of M87 ($r > r_a$) showed that $(M/L)_* = 6$. Moreover, inside r_a there was no spectroscopic evidence for a dramatic change in the stellar population. Accordingly, Sargent et al. set the second term in brackets equal to zero; the other two terms are directly observed.

In particular, for $r = 1\rlap{.}''5$ it was found that $-d\ln L_*/d\ln r = 1.0 \pm 0.1$ and $-d\ln\sigma_v^2/d\ln r = 0.6 \pm 0.3$. With these values it was found that

$$M(r < 1.5'') = 6.5 \pm 1.5 \times 10^9 M_\odot \quad (11)$$

while the photometric measurements give

$$L(r < 1.5'') = 11.2 \pm 0.8 \times 10^7 L_\odot \quad (12)$$

The resulting *total* mass-to-light ratio $\langle M/L\rangle (r < 1.5'') = 58 \pm 16$, considerably larger than that obtained for the stellar population, $M/L = 6$. Thus, Sargent et al. concluded that M87 contains a large central dark mass $M \sim 5 \times 10^9 M_\odot$, a result based on dynamical arguments which agreed with that obtained by Young et al. (1978) on purely photometric evidence. There is, of course, no compelling direct evidence that the central mass, if it exists, is a black hole. On the other hand, the spectra obtained by Sargent et al. (1978) showed no evidence for any change in the stellar population of M87 inside r_a.

Later Young et al. (1979)[10] found a photometric anomaly in the light profile of the radio galaxy NGC 6251 which again was interpreted in terms of a central point mass $M \sim 5 \times 10^9 M_\odot$. This galaxy is about 6 times more distant than M87 and no complementary spectroscopic measurements have been reported.

THEORETICAL PROBLEMS

Duncan and Wheeler (1980)[11] suggested that the Young et al. (1978) and Sargent et al. (1978) observations of M87 could be explained without recourse to a central black hole if the assumption that the velocity dispersion is isotropic was dropped. Thus, in spherical polar coordinates the velocity dispersion may be written in terms of its components as

$$\sigma^2 = \sigma_r^2 + \sigma_\theta^2 + \sigma_\phi^2. \quad (13)$$

We define the transverse component σ_T such that $\sigma_T^2 = \sigma_\theta^2 + \sigma_\phi^2$. Then if σ is isotropic $\sigma_r^2 = \sigma_theta^2 = \sigma_\phi^2$ and $\sigma_T^2 = 2\sigma_r^2$.

In the case of an anisotropic velocity ellipsoid, the first moment of the Boltzmann equation becomes

$$\frac{d}{dr}\left\{\rho(r)\theta_v^2\right\} = -\frac{GM(v)}{r^2}\rho_*(r) - \rho_*(r)\left\{\frac{2\sigma_v^2 - \sigma_T^2}{r}\right\} \quad (14)$$

With a suitable choice of σ_T/σ_r Duncan and Wheeler were able to fit the observations of Young *et al.* and Sargent *et al.* with a central light spike (presumed to be non-thermal) and no central point mass. The resulting model has $\sigma_T = 0$ in the center (by symmetry) and at large r and a variable M/L ratio.

Binney and Mamon (1982)[12] developed a technique for determining if the observed $L(r)$ and $\sigma_v(r)$ are consistent with a constant M/L ratio. The method then allows one to determine the required anisotropy in the velocity ellipsoid as a function of r. They introduced an "anisotropy parameter" $\beta(r)$ defined such that

$$\beta(r) = \frac{2\sigma_r^2 - \sigma_T^2}{2\sigma_r^2} \qquad (15)$$

so that $\beta = 0$ for an isotropic velocity ellipsoid and $\beta = 1$ for one which contains only radial orbits.

A criticism of Binney and Mamon's procedure is that it does not guarantee that a distribution function $f(E,J)$ exists that is consistent with the run of $\beta(r)$ which is obtained. This criticism is not trivial because the M87 observations require an isotropic velocity ellipsoid in the center and in the outer parts of the galaxy, with a highly radial velocity ellipsoid just outside the nucleus.

Accordingly, Newton and Binney (1984)[13] developed an algorithm for constructing a non–negative distribution function $f(E,J)$ that generates given runs of $L(r)$ and $\sigma_v(r)$ in an adopted gravitational potential Φ. (Of course, $\Phi(r)$ can be obtained from $L(r)$ if M/L is assumed constant.) They showed that a distribution function could be obtained which reproduces the M87 observations assuming a constant M/L ratio. Like the previous models, the successful one has a very large ratio σ_r/σ_T close to the nucleus.

Thus, until very recently, it seemed that the M87 observations could be explained without invoking a central black hole. Recently, however, doubts have been cast on this conclusion. Phinney (1985)[14] and Merritt (1986)[15] have shown by numerical simulations that spherically symmetric galaxies with a large ratio σ_r/σ_T are unstable to the formation of a bar (which is, of course, non-rotating). Phinney's models are arranged to have the same initial light distribution as a King model; a bar then forms on a short time scale by a process which Phinney envisages to be similar to a Jeans instability.

Since accurately measured isophotes of M87 show no signs of a bar–like structure, it seems likely that we cannot appeal to an anisotropic velocity distribution to explain the observations. Thus, it appears that the anomalies inside the core nucleus are best explained by invoking a central mass concentration; however, is this a black hole or a massive star cluster?

OBSERVATIONAL PROBLEMS

It was noted earlier that the velocity dispersion measurements of Sargent *et al.* (1978) had a resolution in the radial direction of only 5".4. A clear prediction of the central black hole hypothesis is that the velocity dispersion should rise with decreasing r, roughly as $\sigma_r \sim r^{-1/2}$. Accordingly, Dressler (1980)[16] obtained spectra of the center of M87 from the Las Campanas Observatory in Chile through a 1"× 1" aperture. The "seeing" was estimated to be 0".75 so that Dressler claimed that his measurement pertained to a radius $r = 0''.37$. Dressler found a velocity dispersion $\sigma_v = 360 \pm 30$ km s^{-1} at this radius from the Mg "b" lines and the Na I "D" lines. From the strengths of these lines Dressler inferred that there was no evidence for a significant non-thermal contribution to the spectrum of the central light spike. Dressler made the important suggestion that the central spike is, in fact, a dense star cluster like that found by Light, Danielson, and Schwarzschild (1974)[17] in the center of M31. The mass M_c and mass-to-light ratio M_c/L_c of this cluster depend on its size. Dressler pointed out that a massive black hole ($M_H \sim 5 \times 10^9 M_\odot$) is compatible with the observations if the cluster is large ($R_{eff} \sim 20$ pc or 0".25) but that ruled out if the cluster is small ($R_{eff} \leq 10$pc or 0".125).

At the same time of Dressler's observation I raised with him the question of how atmospheric refraction had affected the location of his 1"×1" aperture; this problem has been discussed recently in the literature by Filippenko (1987)[18]. At the latitude of Las Campanas M87 never rises above an air mass of 1.33; Filippenko's calculations show that at this optimum air mass the image at $\lambda 3500$ is displaced from that at $\lambda 5000$ by 1".01, while the image at $\lambda 6500$ is displaced in the opposite direction by $-0".38$. In fact, the total extent of the refracted image from the blue end of Dressler's observed range ($\lambda 3300$) to the red end ($\lambda 7000$) is 1".72 (Filippenko 1987). Thus, it was not possible for him to have centered the 1"× 1" aperture on the nucleus at all wavelengths. Dressler believed that during his observation the M87 light spike was centered at around $\lambda 5500$ because that is the wavelength at which the television guiding camera is most sensitive. However, this is in conflict with Dressler's (1980) own observation that the Ca II H and K absorption lines and the G-band are diluted by a substantial ($\sim 30\%$) contribution from a non-stellar continuum. Thus observations would imply that the light spike was centered at $\sim \lambda 4200$ and that it was only partially inside the 1"× 1" aperture at $\lambda 5500$, the central wavelength of the range used for the velocity dispersion determination.

In later observations from Palomar which did not suffer from the above difficulties with atmospheric refraction, Young, Gunn, and Kristian (1980, unpublished) obtained $\sigma_v = 400$ km s^{-1} for the center of M87; this is confirmed by recent measurements by Filippenko and Sargent (1986, unpublished). These last authors obtain $\sigma \approx 400$ km s^{-1} and conclude that the non-stellar light is a significant fraction of the total even at visual wavelengths.

A more recent observational development has been Bagnuolo's (1987)[19] measurement of the diameter of the central light spike in M87 using a series of short exposures at Haleakala in order to "freeze" the seeing. Bagnuolo obtains a full width at half maximum intensity of $\leq 0".25$ at the V-band. The corresponding effective radius is $b = 0".1 = 8$pc. It is not clear how much of the light inside 0".25 ($V = 16.3$ mag, corresponding to $L_v = 4.8 \times 10^7 L_\odot$) is non-thermal.

IMPLICATIONS OF THE RECENT M87 OBSERVATIONS

We shall treat the central spike in M87 as an isolated star cluster–the effect of stars from the rest of the galaxy is negligible. For simplicity we represent the cluster by an $n = 5$ polytrope in which the gravitational potential is given by

$$\Omega = -\frac{3\pi}{32} \frac{GM_c^2}{b} \qquad (16)$$

where M_c is the mass of the cluster and b is the effective radius. Applying the virial theorem

$$-\Omega = 2T = 2M_c \sigma_v^2 \qquad (17)$$

where σ_v is the radial (i.e., measured) component of the velocity dispersion.

The mass of the cluster is then given in terms of measurable quantities by

$$M_c = \frac{32}{\pi} \frac{\sigma_v^2 b}{G}. \qquad (18)$$

If we take Dressler's (1980) value of $\sigma_v = 360$ km s^{-1} and assume a distance of 15 Mpc, so that $1'' = 73$ pc

$$M_c = 3.3 \times 10^8 \left(\frac{b}{1\text{pc}}\right) M_\odot \qquad (19)$$

and the mass-to-light ratio of the cluster is:

$$\frac{M_c}{L_c} = 6.7 \left(\frac{b}{1\text{pc}}\right) M_\odot. \qquad (20)$$

(Note that some of the luminosity is non-thermal.)

The relaxation time of the cluster is

$$t_R = \frac{\pi \sigma_v^2 b^2}{G m_* \ln\Lambda} = 5 \times 10^9 \left(\frac{b}{1\text{pc}}\right) \text{ years} \qquad (21)$$

and the stellar collision time is

$$\frac{t_c}{t_R} = \left(\frac{v_e}{\sigma_v}\right)^4 \frac{\ln\Lambda}{16\sqrt{2}} \left(1 + \frac{v_e^2}{\sigma_v^2}\right)^{-1} \sim 9 \qquad (22)$$

where Λ is the "coulomb cutoff"–the ratio of the radius of the system to the radius of a star, and $v_e \sim 600$ km s^{-1} is the escape speed from the surface of a star.

We see that if the "spike" has $b = 0\rlap{.}''1 = 7$pc, then

$$M_c = 2.4 \times 10^9 M_\odot. \qquad (23)$$

Interestingly, this is the largest mass which is compatible with the photometric anomaly inside r_c discovered by Young et al. (1978). The corresponding mass–to–light ratio is

$$\frac{M_c}{L} = 45 \qquad (24)$$

which is a lower limit for the stellar constituent of the spike if some of its luminosity is non–thermal.

If b is much less than $0\rlap{.}''1$ then t_R and t_c become dangerously short as compared to the Hubble time. Therefore, there are two possible interpretations of the present observations of the central spike:

1. A massive, dense star cluster whose M/L is much higher than that which obtains just outside the nucleus of M87.
2. A star cluster of normal M^*/L^* surrounding a massive black hole.

RECENT WORK ON M31 AND M32

Evidence for a central mass concentration in M32 was found by Tonry (1984)[20] and by Dressler (1984)[21]. This dwarf E galaxy has a central light spike and a centrally peaked distribution of velocity dispersion. Both authors also found that the central regions of M32 are rotating with an amplitude of at least 50 km s^{-1}; outside the center the rotation falls off in a Keplerian fashion. New data on M32 have been discussed by Dressler and Richstone; their implications are examined by Richstone (1987)[22]. Reasonable models require a mass of 1.7 to $2.0 \times 10^7 M_\odot$ inside $1''$ or 3 pc. A constant value of M/L leads to $1.0 \times 10^7 M_\odot$ inside this radius (Tonry 1984). Therefore, M32 either contains a dark mass of at least $5 \times 10^6 M_\odot$ within the central 3 pc or the mass–to–light ratio of the stellar population increases rapidly in the center. Richstone

(1987) has described a specific model which reproduces the observations and which contains a central black hole of mass $M_H = 2 \times 10^7 M_\odot$. More recently, as reported by Richstone (1987), both Dressler and Kormendy have obtained new observations of the central bulge of M31 which are shown to have a very small nucleus by Light, Danielson, and Schwarzschild (1974). Richstone asserts that the rise in velocity dispersion and the rapid rotation of the stars in the center of M31 are currently best modeled by a central dark mass–in this case $M_H = 6 \times 10^7 M_\odot$.

THE NEXT STEPS

We have seen that there is tantalizing, but so far not conclusive, evidence that supermassive black holes exist in the nuclei of several nearby galaxies. Clearly, more theoretical work is required to clarify the question of the stability of anisotropic velocity ellipsoids. On the observational side we still await a really satisfactory measurement of the velocity dispersion in the center of M87 which can be associated with a well–defined radius. Moreover, the question of how much of the central light is non–thermal is still of critical importance. In the future it is clear that observations of both the light profile and the velocity dispersion in the center of M87 from the Hubble Space Telescope will be decisive–although the latter will be a very difficult measurement.

Acknowledgements

I thank the organizers for inviting me to participate in the Townes Symposium. I also thank A. Filippenko and S. Phinney for discussing several aspects of this paper. The work was supported in part by Grant AST84–16704 from the National Science Foundation.

REFERENCES

1. E. E. Salpeter, Astrophys. J. 140, 796 (1964).
2. D. Lynden-Bell, Nature, 223, 690 (1969).
3. I. R. King, Astron. J. 71, 64 (1966).
4. P. J. E. Peebles, Astrophys. J. 178, 371 (1972).
5. J. N. Bahcall and R. A. Wolf, Astrophys. J. 209, 214 (1976)..
6. P. J. Young, Astrophys. J. 217, 287 (1977).
7. W. L. W. Sargent, P. L. Schechter, A. Boskenberg, and K. Shortridge, Astrophys. J. 212, 326 (1977).
8. P. J. Young, J. A. Westphal, J. Kristian, C. P. Wilson, and F. P. Landauer, Astrophys. J. 221, 721 (1978).
9. W. L. W. Sargent, P. J. Young, A. Boksenberg, K. Shortridge, C. R. Lynds, and F. D. A. Hartwick, Astrophys. J. 221, 731 (1978).
10. P. J. Young, W. L. W. Sargent, J. Kristian, and J. A. Westphal, Astrophys. J. 234, 76 (1979).
11. M. J. Duncan and J. C. Wheeler, Astrophys. J. Letters 237, L27 (1980) .
12. J. Binney and G. A. Mamon, Monthly Notices Roy. Astron. Soc. 200, 361 (1982).
13. A. J. Newton and J. Binney, Monthly Notices Roy. Astron. Soc. 210, 711 (1984).
14. S. Phinney, private communication (1985).
15. D. Merritt, preprint (1987).
16. A. Dressler, Astrophys. J. Letters 240, L11 (1980).
17. E. S. Light, R. E. Danielson, and M. Schwarzschild, Astrophys. J. 194, 257 (1974).

18. A. V. Filippenko, in "Supermassive Black Holes", ed. M. Kafatos (Cambridge: Cambridge University Press 1987).
19. W. G. Bagnuolo, Nature (in press) (1987).
20. J. L. Tonry, Astrophys. J. Letters 283, L27 (1984).
21. A. Dressler, Astrophys. J. 286, 97 (1984).
22. D. O. Richstone, in "Supermassive Black Holes", ed. M. Kafatos (Cambridge: Cambridge University Press 1987).

THE CENTRAL OBJECT: SOME COMMENTS AND SPECULATIONS

Martin J. Rees
Institute of Astronomy, Madingley Road,
Cambridge CB3 OHA, England

ABSTRACT

There is strong circumstantial evidence for a $\sim 10^6$ $M_\odot$ black hole at the Galactic Center. Such an object could account naturally for the unusual compact radio source. Capture and destruction of stars by the hole could lead to directional ejection of some of the debris; this phenomenon might be relevant to the energetics and morphology within the central "cavity".

INTRODUCTION

The phenomena at the Galactic Center are still baffling. Theorists lag unusually far behind the observers - testimony to the region's intrinsic complexity and also to how observations have forged ahead under Professor Townes' inspiring leadership. Since the Caltech conference nearly five years ago[1], the kinematics of the clouds have clarified, and the annihilation γ-ray flux has resolutely remained in the 'low' (i.e. unobservable) state. Radio data now offer still better resolution on the compact central source, as well as revealing (on a much larger scale) the extraordinary "loops".

I should like to discuss the hypothesis that our Galactic Center harbours a black hole with the 'advertised' mass of a few million solar masses. My motive is less the (still ambiguous) direct evidence than a belief that black holes are natural — indeed, perhaps inevitable - outcomes of past activity in any galaxy. This was what led Lynden-Bell and myself,[2] back in 1971, to explore the notion of a black hole at our Galactic Center. Recent discoveries by Sargent and others (these proceedings and ref. 3) reveal the near-ubiquity of activity (at some level) in galactic nuclei; in the light of these findings, we would be 'underprivileged' if our own Galaxy were not endowed with a black hole.

I shall focus on the central parsec, ignoring the bewildering complexities of, for instance, the inner disc and radio arc which we have heard about from other speakers. The key question is whether the dynamics in this region are indeed dominated by a $\sim 10^6$ $M_\odot$ black hole. This view has been advocated by the Berkeley group in particular, but I'm sure they would themselves accept that any such inference depends on an interpretation of the gas kinematics that is by no means mandatory; moreover, as the contributions from Drs Allen and Ekers emphasise, any attempt at interpretation is confused by the possibility of substructure within IRS 16, differential reddening along different lines of sight, the imprecise location of the 'dynamical center', and the positional disagreement between the radio and infrared structure.

THE COMPACT NON-THERMAL SOURCE

If a $\sim 10^6 M_\odot$ black hole <u>were</u> present, it would be natural to associate it with the compact radio source discussed by Dr Lo. Indeed low-level accretion onto the hole could then offer a natural interpretation of this apparently-unique source. In the proceedings of the Caltech conference,[4] I argued that a compact radio source resembling the one actually observed would be an almost inevitable manifestation of a large black hole, and I shall 'recap' the subject only briefly here. Inflow must now be proceeding at far below the critical rate: the hole must currently be 'starved' of fuel - a "critical" rate of infall would yield a (predominantly thermal) luminosity of order $L_{Edd} \simeq 10^{44}$ erg s^{-1} from the Galactic Center, which there certainly is not. Only $\sim 10^{-5} M_\odot$ yr^{-1} would suffice to provide the entire 10^{44} erg s^{-1} observed from the central parsec. In terms of this rate $\dot{M}_{-5} = \dot{M}/10^{-5} M_\odot$ yr^{-1}, the field strength corresponding to equipartition with kinetic energy would be

$$B_{eq} \simeq 10^4 \dot{M}_{-5}^{\frac{1}{2}} M_{h6}^{-1} \left(v_{inflow}/v_{freefall}\right)^{-\frac{1}{2}} (r/r_g)^{-\frac{5}{4}} G \quad (1)$$

Shearing motions could amplify a pre-existiing field to a strength $\sim B_{eq}$. If this happens in a flow with low $\dot{M}$, then the simplest outcome is an "ion-supported torus". More realistically, one might expect a multiphase gas around the hole. Some cool material, configured either in a thin disc or in discrete blobs, may be present, radiating thermally in the UV. But there would then also be a physically thick corona in which electrons would radiate by synchrotron or inverse Compton emission. This corona could be supported by magnetic fields anchored in the disc or by pressure of the ions. Because energy transfer from ions to electrons would occur slower than the infall, ions would remain at the virial temperature even if electrons could cool. For low $\dot{M}$, the only efficient radiation processes are synchrotron or Compton. The radio emission would then be due to relativistic electrons (with $\gamma_e \simeq$ 10-100) radiating in a field of $10^2 - 10^4$ G. The source could be gravitationally bound to the black hole - either by ions intermixed with the plasma or by a cool disc or blobs magnetically coupled to the region occupied by the relativistic electrons. The radio source dimensions are then a few hundred times the gravitational radius $r_g = GM/c^2 \simeq 1.5 \times 10^{11}$ ($M_h/10^6 M_\odot$) cm. If the immediate environs of the hole were to produce most of the $10^7 L_\odot$ of optical/UV emission via thermal processes, then the magnetic field could still be strong enough to guarantee that Compton X-rays were not excessive.

In general, we would expect the intrinsic size of the source to decrease with increasing frequency: B_{eq} is higher at small r, whereas self-absorption constraints imply that the lower frequencies must come from a larger region. The synchrotron spectrum would extend up at least into the infrared: for $\gamma_e \simeq 10$ and $B = 3 \times 10^4$ G the optically-thin synchrotron spectrum peaks at $\gtrsim 10^{12}$ Hz. Thus the most clear-cut observational consequences of low-level accretion onto a 10^6 $M_\odot$ hole would be synchrotron-type emission in the radio and far infrared, the latter coming from a source only a few times r_g in size. The Galactic Center would then be the only incoherent source where radio techniques can probe close to a relativistic object.

The linear polarization of the radio source would depend on the field configuration; since the relevant electrons have $\gamma_e \simeq 10$, up to a few per cent circular polarization may be expected. However, this is very much an upper limit, because we would expect cancellations due to different parts of the source having opposite signs for $B_\parallel$, and due also to self-absorption effects. If high circular polarization were observed this would suggest the more radical view that the radio emission was coherent. A "synchrotron maser" is impossible unless the plasma density or particle anisotropy are implausibly contrived. Coherent <u>cyclotron</u> emission, however, arises more readily; in fields $\sim 10^4$ G (cf. (1)) this would be in the GHz band. The brightness temperature would then be $\gtrsim 10^{14}$ K. Such radiation would be subject to spontaneous (and induced) Compton scattering; while this would be catastrophically effective if a <u>quasar's</u> radio luminosity came from dimensions of only a few times r_g, such need not be the case for the less extreme brightness temperatures appropriate in this far less powerful source ($L \lesssim 10^{34}$ erg s^{-1}).

Irrespective of its size, the non-thermal source at the Galactic Center has the rather surprising property that its apparent luminosity over the entire X-ray and γ-ray band — from a few keV up to ~ 1 Gev photon energies - is no larger than the power going into the annihilation line (i.e. into the rest mass of the pairs). This could imply any of the following three possibilities.

(i) <u>The pairs are produced at energies of only $\lesssim 1$ Mev.</u> This could happen if there were a central source emitting primarily γ-rays, which was sufficiently compact that $\gamma + \gamma \to e^+ + e^-$ occurred. The requisite dimensions, for a pair luminosity $\sim 10^{38}$ erg s^{-1}, are $\sim 10^9$ cm. This would be <u>smaller</u> than the gravitational radius of a black hole unless its mass were below a few thousand solar masses[5,6]. Alternatively, the pair may result from radioactive decay implied by the Al^{26} γ-ray line feature.

(ii) <u>The pairs could be produced with high Lorentz factors but could lose their energy (before annihilation) via synchrotron radiation in the form of soft (optical or UV) photons.</u> Even if all the ionizing flux came from this central source — and none from stars — the pairs would still need to form with $\gamma_e \lesssim 10^3$.

(iii) <u>Much more continuum luminosity is perhaps being beamed along other directions than our line of sight.</u> An elaborate model along these lines - the "gamma gun" - has been developed by Kardashev and Novikov[7].

If there were no central masses larger than $10^3 - 10^4$ $M_\odot$, then the radio source could not be gravitationally bound, but might instead come from a wind[8]. While the compact radio source does not unambiguously require a massive black hole, the presence of such an object, accreting at a low rate, would almost inevitably yield radio emission - the <u>absence</u> of a peculiar radio source at the Galactic Center would therefore have been evidence <u>against</u> a massive black hole there.

STELLAR CAPTURE AND DISRUPTION

Despite the recent observational developments, I honestly do not think the net strength of the case for a massive central black hole is very different from what it was five years ago. The compact radio source is still the most unambiguous clue that the Galactic Center contains some kind of body that is unique within our galaxy; but what is really needed is some phenomenon that cannot be explained in any other way.

An inevitable concomitant of a black hole's presence would be tidal capture and disruption of stars. The consequences of this process are not necessarily inconsistent with what we observe; more speculatively, I wish to suggest (cf. ref. 9) that we may already have evidence for it.

Whereas the gravitational radius r_g scales with the hole's mass, the tidal radius r_T (defined as the radius within which a star of given type would be tidally disrupted) grows only as $M^{1/3}$. A hole exceeding $10^8 M_\odot$ would swallow solar-type stars without disrupting them (a fact whose relevance to quasar models was first emphasised by Hills[10]). But a hole within the mass range relevant to our Galactic Center, would tidally disrupt all stars of density $\lesssim 10^4$ gm cm^{-3} before swallowing them. The fate of the debris from such stars is then crucially important.

Ozernoi and his collaborators have, in a series of interesting papers,[11-13] highlighted the importance of stellar capture at the Galactic Center. The interval between successive events of this kind, which occur whenever a star passes within a distance r_T of the hole, was estimated by Gurzadyan and Ozernoi[13]; for a star density of $10^5 M_\odot$ pc^{-3} within the central pc, and an assumed velocity dispersion of 200 km s^{-1} it would be $\sim 10^4 M_{h6}^{-4/3}$ yrs.

This estimate is based on assuming a star density of 10^5 pc^{-3} within the central parsec (which would not of course be obligatory if the radiation was all energised by a black hole); the capture rate, and its dependence on the 4/3 power of M_h, would both, in any case, be modified by loss-cone effects within the 'cusp' of stars in orbits gravitationally-bound to the hole. Such refinements are, however, minor compared to the uncertainties about the fate of the debris after a star has been torn apart. If all the debris were swallowed, and the resultant release of gravitational binding energy achieved the efficiency (10 per cent) characteristic of steady thin accretion discs, then the mean accretion-powered luminosity would be $5 \times 10^{41} M_{h6}^{4/3}$ erg s^{-1}.

The observational consequences depend, however, on the uncertain answers to several interlinked questions: What fraction of the debris actually goes down the hole, rather than being expelled? What is the radiative efficiency for the accretion process? (In other words, how many ergs of energy are radiated for each gram that is swallowed?) How long does it take to "digest" one star? In particular, how does the "flare duration" and decay timescale for such a process compare with the interval between one stellar disruption and the next?

If a solar-type star approaching on an almost "parabolic" orbit with small impact parameter is disrupted by passage within a distance $r_T (\sim 10^2 r_g)$ of a $\sim 10^6 M_\odot$ black hole, the bits of debris move out along orbits whose mean binding energy (to the hole) is $GM_*/r_* \simeq 10^{-5} c^2$. This is because the energy needed to tear the star apart has come from the incoming orbital energy. These orbits are very eccentric with typical major axes $\sim 10^5 r_g$; the "mean orbit" for the debris has a period of $\sim 30\, M_{h6}$ yrs (t_{orb}).

Three mechanisms may be distinguished whereby some of the debris may nonetheless escape:

(a) Debris that ends up bound to the hole, on a range of elliptical orbits with specific binding energy $\sim 10^{-5} c^2$, would form an axisymmetric torus after very few orbits. But such a torus would be likely to have such high viscosity that energy would be released at far above the Eddington limit. This would result in much ejection. The fraction of the stellar rest mass energy emitted as radiation depends on the fraction swallowed (rather than expelled) and on the efficiency with which it can radiate before being swallowed. However, all the action would be over within a few times t_{orb}. We cannot rule out $\sim 10^{44}$ erg s^{-1} provided that it is a flare like event with a short duty cycle (i.e. a duration $\ll 10^4$ yrs). The expulsion driven by radiation pressure would be anisotropic, probably in a twin jet configuration aligned with the angular momentum vector of the original stellar orbit, and with ejection velocities up to $\sim 0.1c$.

(b) Since the original binding energy of the disrupted star must be supplied by its orbital energy, the debris must <u>on average</u> be bound to the hole by an energy $\sim v_*^2$ per unit mass, v_* being the escape velocity from the surface of the star. However as Lacy et al.[14] emphasized, some fraction of the debris can escape at $\gtrsim 1000$ km s^{-1}. The reason for this is that at peribothron a star undergoing tidal disruption is moving at $V \simeq 3 \times 10^4 M_{h6}^{1/3}$ km s^{-1}: it becomes somewhat compressed and elongated into a prolate shape (cf. Carter and Luminet[15,16]) and pressure gradients can impart to material on the leading side of the star an <u>excess</u> velocity δv over the parabolic orbital velocity which is a <u>significant</u> fraction of v_*. This corresponds to a large excess orbital energy - enough for the <u>debris</u> to escape on a hyperbolic orbit with terminal velocity $\sim \sqrt{(\delta v) v_*}$. Whenever a star is disrupted, some fraction of its "remains" will spray out in a fan or cone.

(c) Stars that penetrate well inside r_T are severely compressed when they pass close to the hole. There is then the possibility of <u>explosive</u> energy release.[15,16] The p-p reaction is too slow to release much energy on a dynamical timescale; however, proton capture on C, N, and O can yield (for solar abundances) enough energy to unbind the star. <u>Most</u> of the debris could then escape on hyperbolic orbits. In the rarer cases when the star captured is a C-burning giant, the compression can lead to an explosive increase in reaction rates.

Every few thousand years, whenever a star passes close enough to the central hole to be disrupted, the Galactic Center would expel a 'ballistic' jet with mass 0.1 -1 $M_\odot$. There are three possibilities:

(i) a double jet, associated with a supercritical torus; the velocity here may be $\sim 0.1c$.

(ii) a fan-like jet, being the part of a tidally-disrupted star expelled on hyperbolic orbits, with speed $\sim 10^3$ km/sec.

(iii) A nuclear powered fan jet, containing possibly almost the entire mass of a star that has passed close enough to the hole for nuclear reactions to release more than $G M_*^2/r_*$ of energy.

CONDITIONS WITHIN THE CENTRAL 1-2 pc: THE CAVITY AND THE ARMS

The fate of fast-moving ejecta, and their observational consequences, would depend on what the surroundings are like. The apparent emptiness if the central 1-2 pc is well known. I suggest that this region is filled with gas with roughly the following properties: $n = 100$ cm^{-3}, $T \simeq 3.10^6$ K. (The corresponding mass within a radius r would then be 20 r_{pc}^3 $M_\odot$). Such a gas would emit soft X-ray luminosity $10^{37} r_{pc}^3$ ergs s^{-1}, which would of course be absorbed in the surrounding few parsecs. (Watson et al's[17] upper limit of 40 cm^{-3} assumed T =40000 K, and can be relaxed if the temperature is higher). Gas with these properties would have a cooling timescale $t_{cool} = 3 \times 10^4$ yr (to be compared with a dynamical or sound-crossing timescale of $t_{dyn} \simeq 5 \times 10^3$ r_{pc} yrs).

The situation would resemble the popular two-phase models for the emission line regions in active galactic nuclei. The gas revealed by the Ne line and by the thermal radio continuum, with temperature $10^3 - 10^4$ K, would then be in pressure balance. Why, then, should it be concentrated in the peculiar 'arm-like' features? One carefully-studied idea is that these arms are streams of tidally-disrupted infalling material, originating in the 2 pc ring.[18,19] I wish to outline an opposite view - that, even though the clouds may be moving inward, their formation was triggered by sporadic rapid outflow from the Centre associated with stellar-disruption events.

Speculation 1

Could the 'spiral' features be vapor trails produced by ejecta from disrupted stars? These features each involve several solar masses of material, moving with speeds ~ 100 km s^{-1}; they are therefore not themselves composed of the high-speed ejecta. Nevertheless, their formation may have been <u>induced</u> by ejecta from disrupted stars.

Suppose that the region within ~ 2 pc is indeed pervaded by hot ionized gas with density $\sim 10^2$ cm^{-3}, as suggested above. [The emission measure from this gas (which may be deficient in dust) would be swamped by that from the $\sim 10^5$ cm^{-3} concentrations within it.] Suppose - as an illustrative example - that each stellar disruption leads to the expulsion of ~ 0.1 $M_\odot$ of material with velocities $\sim 10^3$ km s^{-1} directed on a cone of solid angle $\sim 10^{-2}$ radians. By the time this debris has travelled ~ 2 pc through gas of density $\sim 10^2$ cm^{-3} it will have swept up at least its own mass. The initial kinetic energy will thus have been turned into thermal energy, leaving an overpressured cone or swath of hotter $\gg 10^6$ K) gas along the path of the ejecta. The clouds of density $\gtrsim 10^4$ cm^{-3} which delineate the observed pseudo-spiral features would then be produced as this overpressured material expands sideways, causing a radiative shock.

Note that in this picture, the observed clouds would consist not of the fast-moving ejecta, but of gas already present in the 'cavity' which has cooled down from $\sim 10^6$ K. Their kinematics would be roughly those of the ambient gas rather than of the ejecta. The details would depend on which of the processes (i) - (iii) gave rise to the ejecta. Conceivably all the three features we now observe result from the most recently swallowed star; the curved arm from mechanism (ii) or (iii) and the two symmetric features (the "bar") from the torus that forms from the bound ejecta (mechanism (i)).

Speculation II

The ejecta from successive disruptions might be responsible for maintaining the central 'cavity'. The stellar debris would contribute to the energy budget within this volume, if not to the mass supply. The mass of fast-moving ejecta at any one time would be small, but could, because of its high density, be conspicuous. It would be interesting to consider whether this might be related to the fast-moving He feature. (Note that the stellar ejecta would be enriched in helium.)

Speculation III

Had one of the recently-disrupted stars been a giant which passed close enough to the hole to trigger explosive C-burning, more drastic abundance anomalies than a mere overabundance of He might arise. It would be interesting to explore the possible relevance of this to the Al^{26} γ-ray emission, which seems to be associated with some special type of event near the Galactic Center, producing this isotope which decays with a lifetime $\sim 10^6$ years.

To give some illustrative numbers: suppose there were 10^5 stars each of 3 M$_\odot$ within the central 2 pc. Such stars spend about 3 per cent of their lives as C-burning giants. There would therefore be 3000 C-burning giants. The question then is whether enough of these would be captured per 10^6 years (the number being enhanced by the larger capture cross section for giants), to produce the 4 M$_\odot$ of Al^{26} whose presence is impied by the observed strength of the γ-ray line. This possibility offers a further motivation for detailed study of the tidal disruption process. Intersting nuclear reactions are expected only for stars that penetrate well inside r_T; they are not expected for the $\gtrsim 10^8$ M$_\odot$

holes associated with powerful AGNs (and their consequences would then in any case be swamped by accretion and/or electromagnetic processes).

It should be clear that these three suggestions are increasingly speculative and 'under-baked'; I advance them with diminishing confidence. So it may be prudent, before the web of conjecture gets stretched too fine, to conclude with the remark that the best evidence in favour of a massive black hole at the Galactic Center is that there is still no firm argument against it.

I am grateful to the organisers of this meeting for offering the opportunity to participate, and to Roger Blandford and Sterl Phinney for help with the speculations.

REFERENCES

1. G.R. Riegler and R.D. Blandford (eds), "The Galactic Center" AIP Conference Proceedings No. 83 (1982)
2. D. Lynden-Bell and M.J. Rees, MNRAS, 152, 461 (1971).
3. A.V. Fillipenko and W.L.W. Sargent, Ap. J. Suppl., 57, 503 (1985).
4. M.J. Rees in ref. 1, p.166.
5. L.M. Ozernoi, these proceedings.
6. D. Eichler and E.S. Phinney (unpublished).
7. N.S. Kardashev, I.D. Novikov, A.G. Polnarev and B.E. Stern on "Positron-Electron Pairs in Astrophysics" AIP Conference Proceedings, eds. M.L. Burns et al. p.253 (1983).
8. S.P. Reynolds and C.F. McKee, Ap. J., 239, 893 (1980).
9. M.J. Rees in "The Milky Way Galaxy", eds H. van Woerden et al. (Reidel, Dordrecht 1985).
10. J.G. Hills, Nature, 254, 295 (1975).
11. L.M. Ozernoi in "Large-Scale Charactristics of the Galaxy" ed W.B. Burton p.395 (Reidel, Dordrecht 1979).
12. V.G. Gurzadyan and L.M. Ozernoi, Astron. Astrophys. 86, 315 (1980).
13. V.G. Gurzadyan and L.M. Ozernoi, Astron. Astrophys. 95, 39 (1981).
14. J.H. Lacy, C.H. Townes and D.J. Hollenbach, Ap. J., 262, 120 (1982).
15. B. Carter and J.P. Luminet, Nature, 296, 211 (1982).
16. B. Carter and J.P. Luminet, MNRAS, 212, 23 (1985) and references cited therein.
17. D.M. Watson, J.W.V. Storey, C.H. Townes and E.E. Haller, Ap. J. (Letters) 241, L.43 (1980).
18. K.Y. Lo and M.J. Claussen, Nature, 306, 647 (1983).
19. P.J. Quinn and G.J. Sussman, Ap. J. 288, 377 (1985).

THE STELLAR POPULATION AT THE GALACTIC CENTER

M. J. Lebofsky and G. H. Rieke
Steward Observatory, University of Arizona, Tucson, Az. 85721

INTRODUCTION

Past studies have indicated that the stellar population at the Galactic Center includes a relatively young component with Source 7 being perhaps no older than 40 million years.[1] The extent of the young population could not be easily judged on the basis of older work. Preliminary results from a 2um survey covering an area of 400 square parsecs (5'x 5') of the Galactic Center are presented here. That a large amount of star formation has occurred recently is confirmed.

DATA

Maps at H and K were obtained using the Steward Observatory 1.54-meter telescope and an infrared camera using a 32x32 HgCdTe hybrid CCD fabricated by Rockwell International. This system was equipped with relay optics yielding a linear scale of 1".2 per pixel. An offset guider was used to position each frame with respect to its neighbors. The observations were calibrated by measurements of standard stars with the same instrument. The entire area mapped required 9x9 frames of 32x32 pixels each; the sensitivity yielded 5-sigma detections at K=13.5 and at H=13.8.

DISCUSSION

The K map shows approximately 1400 stars, corresponding to an average of 1 star per 40 pixel areas. Because of strong foreground extinction in certain areas of the map, the stars are not uniformly distributed; over about half the area, the map is confusion limited at a K magnitude of 13.5.

For those objects detected at H and K both, the extinction was computed assuming an intrinsic H-K=0.3. See Figure 1 for the distribution of observed H-K colors which translates almost directly into extinction. The average value for all objects is A_v=30 with a few objects having A_v as high as 70. More heavily reddened objects would not have been detected at H with the integration times used.

Figure 2 shows the dereddened K magnitudes assuming an intrinsic H-K=0.3. By using an approximate conversion from work by Frogel and Whitford[3] for late-type stars in Baade's Window, these K magnitudes can be converted to bolometric magnitudes, assuming a distance of 7 Kpc. Figure 3 shows the distribution of bolometric magnitudes for Galactic Center stars and for Baade's Window stars from Frogel and WHitford[3]. The Galactic Center objects appear to represent a stellar population about 3 magnitudes brighter than seen in Baade's window. This result would be expected if there has been recent (within the last 100 million years) star formation at the Galactic Center.

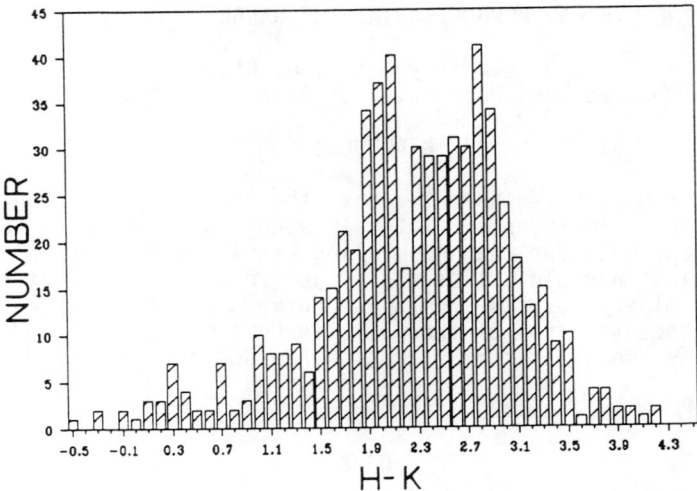

Figure 1: The distribution of observed H-K colors at the Galactic Center.

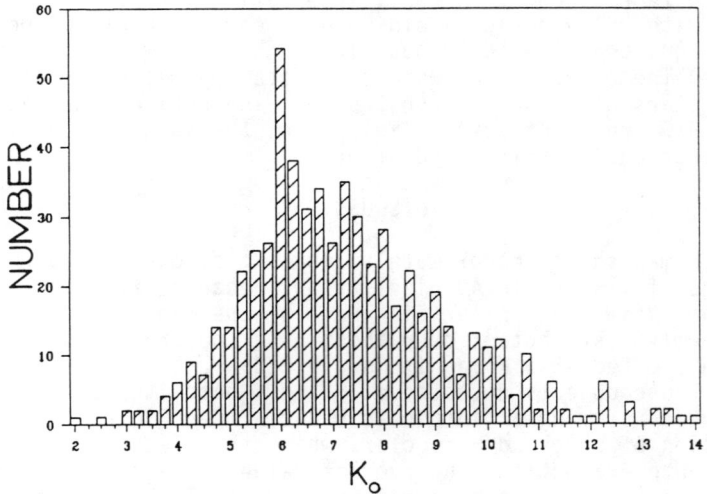

Figure 2: The distribution of dereddened K magnitudes.

Further evidence for the Galactic Center stars forming a separate population from the Galactic bulge comes from Figure 4. Figure 4a. shows the distribution of flux for unresolved sources averaged across each frame. Figure 4b. shows the underlying, diffuse flux which is presumably from the bulge. Neither set of fluxes have been dereddened. Figure 4c. shows the ratio of the two fluxes. The unresolved source flux does not track the underlying bulge.

The other possibility is that this group of stars are foreground objects, and represent members of the same population as

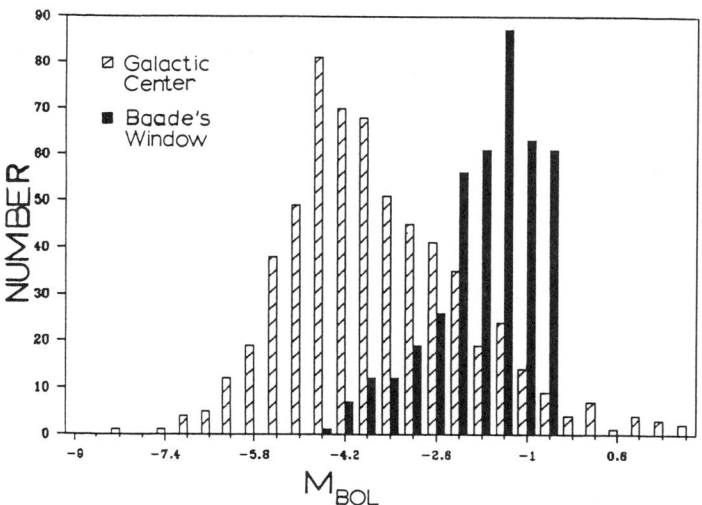

Figure 3: The distribution of bolometric magnitudes assuming a distance of 7 Kpc.

seen in Baade's Window. If so, these objects must lie 4 times closer to compensate for their greater brightness. To estimate whether this is probable, the surface brigthness at 2.4um from Oda et al.[4] for a galactic longitude corresponding to 5.4 Kpc from the center (l=38°) is converted to an equivalent number of stars per square degree. This number of stars is computed using the luminosity function in

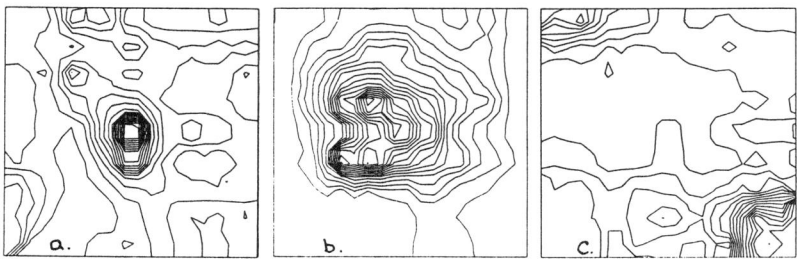

Figure 4: a.) Flux distribution for unresolved sources, b). Flux distribution for diffuse sources, c.) ratio of unresolved to diffuse flux.

Frogel and Whitford[3], and hence is an overestimate of the number of stars present brighter than the Frogel and Whitford completeness limit. The number of stars at this position is less than 6000/square degree brighter than $K_o=7.5$. The areal density on the Galactic Center map corresponds to 57000/square degree brighter than $K_o=7.5$ which implies that these stars cannot be interpreted as just foreground bulge stars. Also note that Oda et al.[4] show that the extinction peaks at the Galactic Center, and that $A_v=30$ is very unlikely for a region at longitude $38°$.

Help from Earl Montgomery is gratefully acknowledged as is support from the National Science Foundation.

REFERENCES

1. Lebofsky, M. J., Rieke, G. H., and Tokunaga, A. T., 1982, ApJ, <u>263</u>, 736.
2. Lebofsky, M. J., Montgomery, E. F., and Kailey, W. F., 1985, in <u>Second Infrared Detector Technology Workshop</u>, ed. C. R. McCreight, (NASA Ames Research Center: Mountain View).
3. Frogel, J. A., and Whitford, A. E., 1986, ApJ, submitted.
4. Oda, N., Maihara, T., Sugiyama, T., and Okuda, H., 1979, A and A, <u>72</u>, 309.

KINEMATICS OF INDIVIDUAL STARS IN THE GALACTIC CENTER

K. Sellgren and D. N. B. Hall
Institute for Astronomy, University of Hawaii, Honolulu, HI 96822

S. G. Kleinmann
University of Massachusetts, Amherst, MA 01003

N. Z. Scoville
California Institute of Technology, Pasadena, CA 91125

ABSTRACT

We have obtained high-resolution 2-μm spectra of six late-type stars within 2 pc of the Galactic Center. We have derived spectral types, reddenings, and radial velocities for these stars. We find the supergiant density in the Galactic Center is lower than in previous determinations. No correspondence is found between stellar velocities and Ne II velocities observed along the same line of sight. No systematic rotation or ordered motion is seen in the stellar velocity distribution. The mass distribution derived from the stellar velocities is compared to those derived from gas velocities and from the 2-μm light.

INTRODUCTION

The Galactic Center provides a unique opportunity to study the mass distribution of a galactic nucleus at high spatial resolution. Previous studies[1] of the mass distribution have used gas velocities, but they can be affected by nongravitational forces. The stars, in contrast, are influenced only by gravity and should more accurately reflect the true mass distribution. We present radial velocity measurements of individual stars in the Galactic Center, obtained to study their kinematics and derive a mass distribution.

STELLAR CHARACTERISTICS

We have obtained 2-μm spectra with 120 km s^{-1} resolution of six stars within 2 pc of the Galactic Center with the Fourier Transform Spectrometer at the Kitt Peak 4-m telescope. The spectra are dominated by CO absorption at 2.3 μm. H_2O absorption is also seen at the blue and red ends of the K band in most sources. The presence of both CO and H_2O absorption, which depend differently on stellar temperature and luminosity, allows us to determine the spectral type and luminosity class of these stars. Table I gives the spectral type and reddening of each star. No spectral type is given for IRS 12, as there appeared to be two sources in our beam. We find that all of the stars except IRS 7 are giants rather than supergiants, in contrast to previous observations[2] finding these stars all to be supergiants. We conclude that while star formation is present in the Galactic Center, as evidenced by the M supergiant IRS 7, the supergiant density may not be as high as previously thought.

Table I Stellar characteristics

Source	Spectral Type	A_K
IRS 7	M2I	3.1 ± 0.2
IRS 11	M5III	2.6 ± 0.2
IRS 12	--	3.0 ± 0.4
IRS 19	M5-7III	2.9 ± 0.3
IRS 22	M5-7III	1.8 ± 0.3
IRS 23	>M7III	3.7 ± 0.9

STELLAR RADIAL VELOCITIES

We have measured the radial velocities of the Galactic Center stars to 10 km s^{-1} by cross-correlating them with late-type stars of known radial velocity. No correspondence is found between the stellar velocities and the Ne II velocities observed along the same line of sight. The velocity structure due to the stars shows no systematic pattern or rotation, unlike the neutral and ionized gas. We show this in Fig. 1, in which the velocities of 2-µm sources (asterisks) and OH/IR stars[3] (crosses) in the Galactic Center are plotted on a 2-cm map of the Galactic Center.

THE MASS DISTRIBUTION DERIVED FROM STELLAR KINEMATICS

We have used the velocities determined for our six 2-µm sources and for seven OH/IR stars[3] to determine the distribution of enclosed mass in the Galactic Center versus projected radius from IRS 16 for a variety of models (Fig. 2). The lower limits are derived from circular Keplerian rotation using individual stellar velocities. The filled circles are a projected mass estimator[4] for an isotropic velocity distribution derived by using the individual stellar velocities grouped into four radius bins. The open circles are from the stellar hydrodynamic equation[5] derived by using the velocity dispersion (80 km s^{-1}) of all 13 stars and by assuming no rotation, no gradient in the velocity dispersion, a power-law index of 1.7 for the radial distribution of starlight, and an isotropic velocity distribution. The diagonal line is a model for the stellar mass distribution of the Galactic Center[6] derived from observations of the 2-µm light distribution. The mass distribution we derive from the stellar velocities is consistent, within the uncertainties, with the mass distribution derived from the stellar light. Our uncertainties in the mass distribution, however, are large, due to the small number of stars measured. This fact is emphasized in Fig. 3, which shows a comparison of the enclosed mass in the Galactic Center, as derived from gas[1] and stellar velocities. The shaded line is a model for the stellar mass distribution derived from the 2-µm light. The two mass distributions derived from the stellar and gas velocities are almost consistent within the error bars, despite the gas kinematics favoring a massive compact object, while the stellar kinematics are consistent with the mass distribution of the stellar cluster. Observations of a larger number of stars will be needed to better determine the mass distribution of the Galactic Center.

REFERENCES

1. M. K. Crawford, R. Genzel, A. I. Harris, D. T. Jaffe, J. H. Lacy, J. B. Lugten, E. Serabyn, and C. H. Townes, Nature 315, 467 (1985).
2. M. J. Lebofsky, G. H. Rieke, and A. T. Tokunaga, Astrophys. J. 263, 736 (1982).
3. A. Winnberg, B. Baud, H. E. Matthews, H. J. Habing, and F. M. Olnon, Astrophys. J. (Letters) 291, L45 (1985).
4. J. N. Bahcall and S. Tremaine, Astrophys. J. 244, 805 (1981).
5. F. D. A. Hartwick and W. L. W. Sargent, Astrophys. J. 221, 512 (1978).
6. R. H. Sanders and T. Lowinger, Astron. J. 77, 292 (1972).

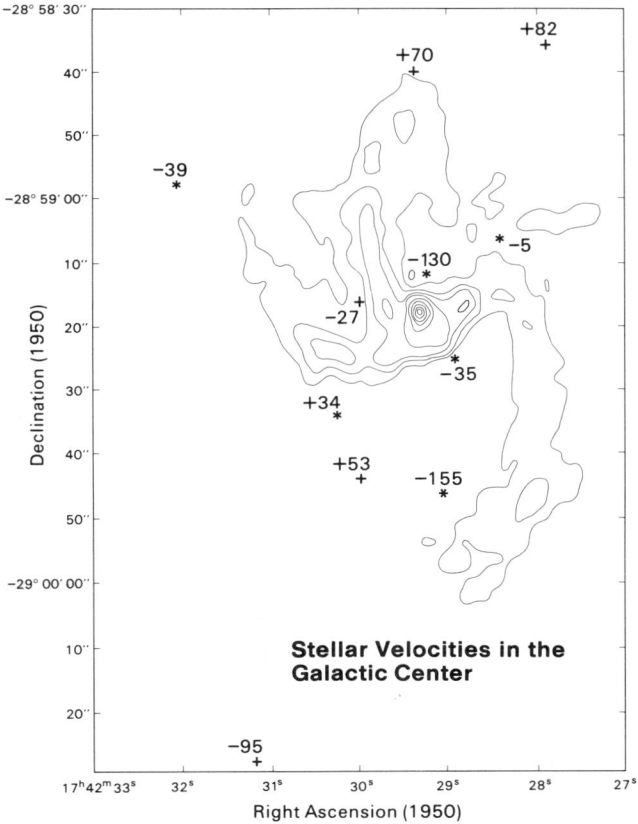

Fig. 1. Stellar velocities in the Galactic Center.

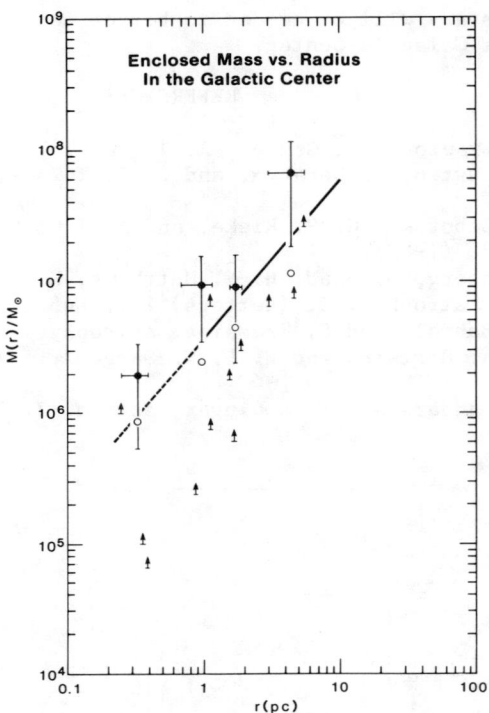

Fig. 2. Mass distribution derived from stellar velocities.

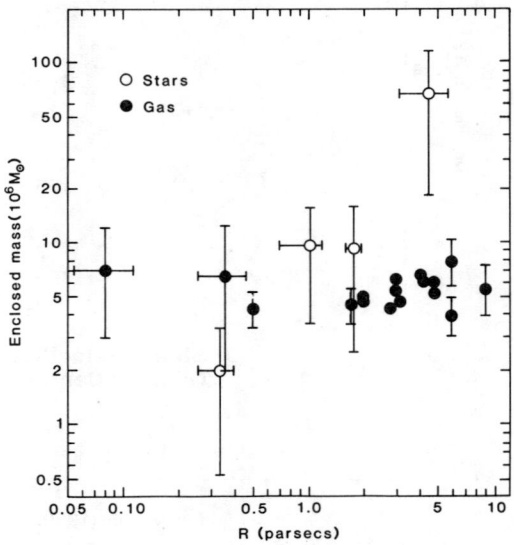

Fig. 3. Mass distributions derived from stellar and gas velocities.

STELLAR KINEMATICS IN THE CENTRAL 10 PC OF THE GALAXY

M. T. McGinn, K. Sellgren, E. E. Becklin, and D. N. B. Hall
Institute for Astronomy, University of Hawaii, Honolulu, HI 96822

I. Gatley
United Kingdom Infrared Telescope, Hilo, HI 96720
National Optical Astronomy Observatories, Tucson, AZ 85726-6732

ABSTRACT

Observations of the profile of the 2.3-μm CO 2-0 bandhead in the Galactic Center are discussed. These large-beam (20" or 1 pc) observations provide information on the integrated stellar velocity and velocity dispersion in the inner 10 pc of the Galaxy. There is evidence that the velocity dispersion dominates the rotational velocity at all positions within 5 pc of the Galactic Center, particularly in the inner regions. We also note a significant difference between the stellar velocities we measure and the gas velocities measured in previous studies.

INTRODUCTION

Observations of the kinematics of the late-type stars in the Galactic Center are critical to determining the central mass distribution. Stars are potentially more accurate tracers of the mass distribution than the gas since they move purely in response to the gravitational field, while the gas may be under the influence of non-gravitational forces. We describe here the experiment we have undertaken to accurately measure the stellar kinematics in the central 10 pc of the Galaxy.

OBSERVATIONS AND ANALYSIS

We have used the Fabry-Perot spectrometer at the 3.8-m UK Infrared Telescope at Mauna Kea Observatory to map the profile of the 2.3-μm CO 2-0 bandhead in a 20" (~1 pc) beam with 120 km s^{-1} resolution. This bandhead is strong in the late-type stars that dominate the 2-μm radiation from the region. Using such a large aperture allows us to determine the integrated stellar velocity and velocity dispersion at each position. The beam positions of our measurements are shown superimposed on a K map of the Galactic Center[1] in Fig. 1.

Throughout the observations, spectra of standard late-type giants of known radial velocity are also recorded. They are compared with the Galactic Center spectra to determine the radial velocity at each position. By convolving these standard star spectra with Gaussian profiles of various widths, it will also be possible to determine the amount of velocity dispersion at each position.

RESULTS

The integrated stellar radial velocities exhibit systematic

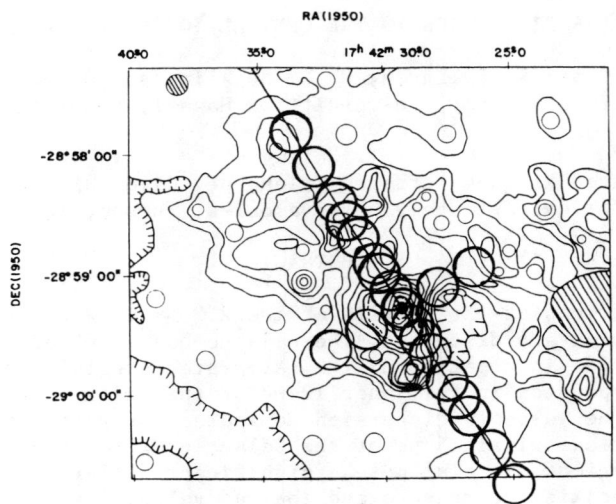

Fig. 1. Beam position of our measurements shown superimposed on a K map of the Galactic Center by Lebofsky.[1] The beam size is 20".

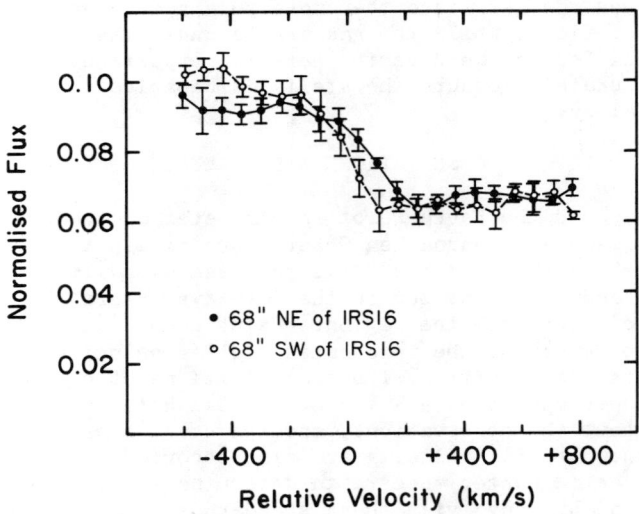

Fig. 2. Spectra of the 2.3-μm CO 2-0 bandhead in two positions in the Galactic plane equidistant from IRS 16. Note the relative velocity shift of 73 ± 30 km s^{-1} in the direction of Galactic rotation.

rotation in the same sense as Galactic rotation. Fig. 2 shows a typical relative velocity shift between two positions in the Galactic plane, equidistant from IRS 16. Using such data, we have plotted rotational velocity as a function of distance from IRS 16 along the Galactic plane. The resultant rotation curve is shown in Fig. 3. Comparison of these data with previous results from studies of the ionized gas[2] indicates that the stellar rotational velocities we measure are lower than the gas velocities. For example, at 2 pc, the gas velocity shift between positions equidistant from IRS 16 is 240 km s^{-1}, while this shift for the stars is 102 km s^{-1}.

The importance of the stellar velocity dispersion in the Galactic Center is clearly seen in Fig. 4, where we have plotted the spectrum of the 2.3-μm CO 2-0 bandhead in a standard M giant and in a Galactic Center position. Note the broadening of the profile in the Galactic Center spectrum. Our results indicate that the velocity dispersion dominates the rotational velocity at all positions, although the ordered rotation becomes more important in the outer regions. The stellar velocity dispersion appears to have a higher value than that derived from the gas motions.

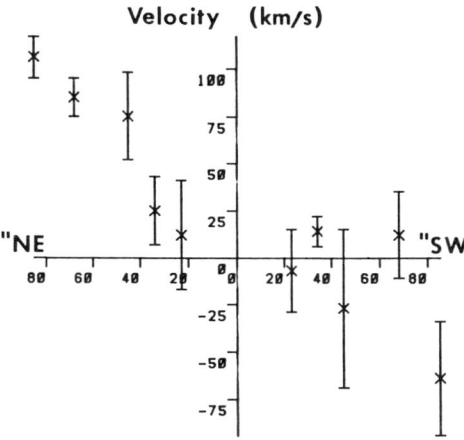

Fig. 3. Rotational velocity as a function of distance from IRS 16 along the Galactic plane, as derived from 2.3-μm observations of the integrated stellar light. All velocities are corrected to LSR.

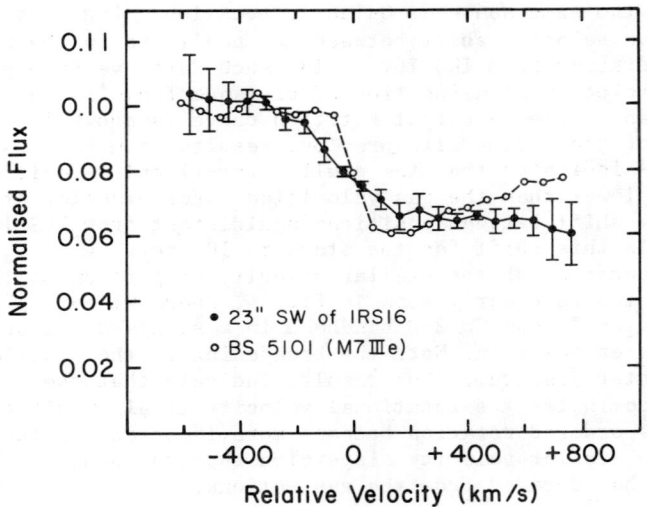

Fig. 4. Spectrum of the 2.3-μm CO 2-0 bandhead in a standard M giant (BS 5101) and in a position 23" SW of IRS 16. Note the broadening of the Galactic Center spectrum, which indicates velocity dispersion.

CONCLUSION

We report preliminary results from a project to extensively map the profile of the stellar CO 2-0 bandhead in the central 10 pc of the Galaxy. Our data will allow us to determine the spatial dependence of the integrated stellar velocity and velocity dispersion. There is an indication of a significant discrepancy between the gas and stellar kinematics and the domination of the stellar rotational velocities by the stellar velocity dispersion, particularly in the inner regions. The ultimate goal of this project is to make an accurate determination of the central mass distribution.

REFERENCES

1. M. J. Lebofsky, Astron. J. 84, 324 (1979).
2. E. Serabyn and J. H. Lacy, Astrophys. J. 293, 445 (1985).

IS THERE A CUSP IN THE STELLAR DISTRIBUTION IN THE GALACTIC CENTER?

G. H. Rieke and M. J. Lebofsky
Steward Observatory, University of Arizona, Tucson, Az. 85721

INTRODUCTION

The cusp in the stellar distribution around the Galactic Center described by Allen, Hyland, and Jones[1] would tend to support arguments from the velocity field for the presence of a black hole. However, the image of a 24 square arcmin field confirms the presence of large nonuniformities in the extinction in this region[2]. The central parsec of the galaxy happens to lie in a minimum of the extinction and furthermore has a number of 2um sources that are not part of the general red giant and supergiant population. These two circumstances can lead to an artificial appearance of a cusp. Moreover, Allen and Sanders[3] suggest that the compact radio source does not coincide with any sufficiently bright near infrared source to be identified with the core of a stellar cusp. However, this argument depends very strongly on the precise location of the compact radio source relative to the infrared maps.

An alternate way to look for a cusp is to measure the surface brightness between the bright stars. Assuming a central density of about 4×10^5 stars pc^{-3} [4], there should be about 4000 stars per square arcsec. Most of these stars will contribute a diffuse, unresolvable background which should be an extremely accurate reflection of the distribution of stellar mass in the region. Searches for a cusp in this diffuse component will also be insensitive to the precise registration of the infrared maps relative to the position of the radio compact source.

MEASUREMENTS

Outside the central parsec, the underlying surface brightness can be measured directly from the frames in the large scale map of Lebofsky and Rieke[2] if regions of high extinction are avoided. We have selected regions which have a relatively large number of bright individual stars with H - K colors indicating extinction greater than A_V = 20 (to avoid foreground stars). Surface brightnesses between the stars were determined relative to surface brightnesses on frames near the edge of the map and where the star counts and colors indicate extremely large extinctions.

Within the central parsec, the underlying surface brightness was determined from a series of 32 X 32 camera frames centered on the compact radio source but with different pixel scales -- 0.25", 0.85", and 1.3". The lowest surface brightness on these frames lies in a small region of very high extinction first noted by Lebofsky[5] about 20" west of the central source complex. This region was used in turn to tie in with the diffuse source measurements for the large scale map. The diffuse source region just outside the central cluster is shown shaded in Figure 1a; the shaded region accounts for roughly 30% of the area of the frame, yet encompasses only 5% of the dynamic range of the frame (after source 7 is excluded; with this

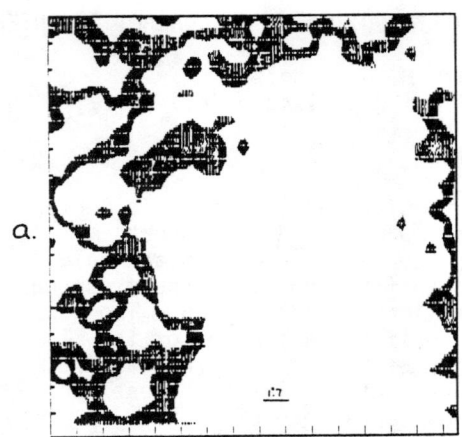 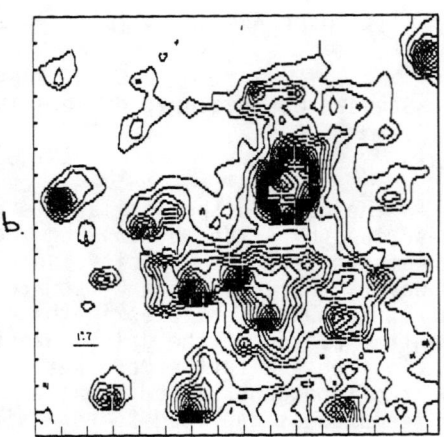

Figure 1: a) The diffuse source region surrounding the central cluster at 2.2um. b) The central cluster at 2.2um. The pixel size is 0.85 arcsec.

source included, it encompasses only 2%). Except for areas around the edges of the large scale frames, which are contiguous to larger areas of heavy extinction, there is only one small area with surface brightness below that of the shaded region; it lies about 11" ESE of source 16 and has a total area of about 6 square arcsec. We interpret this area to be a small cloud of heavy extinction; however, if it is sampling the true underlying surface brightness, the conclusions below are modified significantly. For reference, Figure 1b shows the central cluster.

By chance, there is a "valley" in the source distribution that comes within about 2 arcsec of the compact radio source -- see Figure 2. The bars in this figure show two estimates of the position

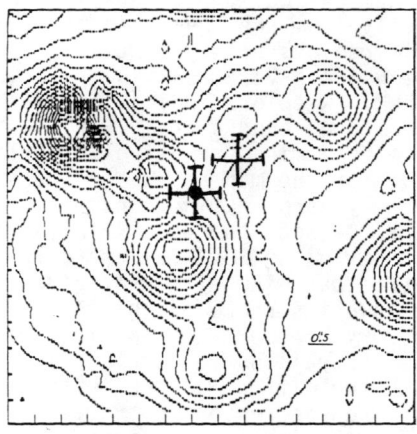

Figure 2: High resolution 2.2um picture of the central cluster. The pixel size is 0.25 arcsec.

of the compact source. The valley is slightly filled in by the wings of the bright source to the SW; this effect has been estimated and corrected from the beam profile on source 7. This corrected surface brightness at the valley floor is used as the estimate of diffuse source brightness at this position.

INTERPRETATION

Figure 3 shows the resulting surface brightness measurements plotted as a function of distance from the compact radio source. The solid line shows a King isothermal stellar cluster model fitted to them. For comparison, Figure 4 is taken from Light, Danielson, and Schwarzschild[6] and shows the surface brightness of the nucleus of M31 to the identical scale. In their figure, the dashed line is the actual data and the solid line is a fit to the data after deconvolving for the beam size.

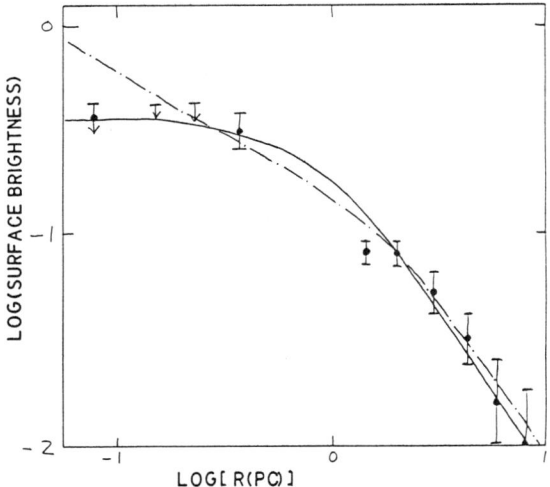

Figure 3: Surface brightness as a function of distance from the compact radio source. The solid line is an isothermal King model while the dash-dot line represents an adiabatically growing black hole model.

There is no evidence in Figure 3 for a cusp near the compact radio source. In a relaxed stellar cluster, a black hole should lead to a cusp within the core diameter that goes as $r^{-3/4}$. However, the two-body interaction timescale for the galactic center is probably too long to achieve this condition. Thus, the nature of any cusp will depend on the history of formation of the black hole and surrounding star cluster. Young[7] has considered the case where a black hole grows adiabatically in an initially isothermal star cluster. His calculations for the case where the hole is three times as massive as the core mass is plotted as a dot-dash line in Figure 3. This case seems to be ruled out by the observations; the resulting upper limit for the hole mass is about 5×10^5 M_o.

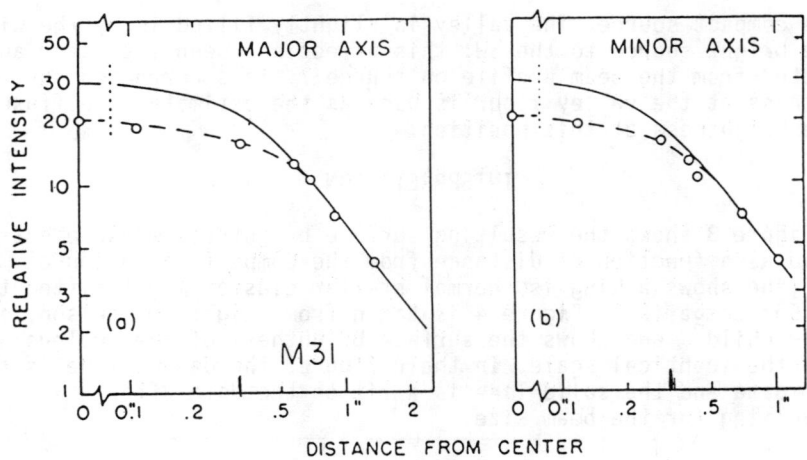

Figure 4: Surface brigthness of the nucleus of M31 with the solid line indicating a fit to the data after deconvolution.

REFERENCES

1. Allen, D. A., Hyland, A. R., and Jones, T. J. 1983, MNRAS, 204, 1145.
2. Lebofsky, M. J., and Rieke, G. H. 1986, this conference.
3. Allen, D. A., and Sanders, R. H. 1986, Nature, 319, 191.
4. Bailey, M. E. 1980, MNRAS, 190, 217.
5. Lebofsky, M. J. 1979, A. J., 84, 324.
6. Light, E. S. Danielson, R. E., and Schwarzschild, M. 1974, ApJ, 194, 257.
7. Young, P. 1980, ApJ, 242, 1232.

THE 18-CM OH DISTRIBUTION IN THE GALACTIC CENTER TORUS

Aage Sandqvist and Roland Karlsson
Stockholm Observatory
S-133 00 Saltsjöbaden, Sweden

John B. Whiteoak and Frank F. Gardner
CSIRO, P.O. Box 76
Epping NSW 2121, Australia

ABSTRACT

The 18-cm OH distribution in the Galactic Center region near Sgr A has been studied in all four of the 1612, 1665, 1667 and 1720 MHz OH lines using the VLA with 4 arcsec angular resolution and 9 km s^{-1} velocity resolution. Three 1667 MHz OH spectral line absorption maps, at +51, +25 and -1 km s^{-1}, covering a 4'.3 x 4'.3 region around Sgr A are presented together with an 18-cm continuum map. In addition, a complete set of velocity maps from +139 to -159 km s^{-1}, covering a 3' x 3' region around the Galactic Center nuclear torus, is presented. Absorption by the +50 and +20 km s^{-1} molecular cloud belt is seen towards Sgr A East, but not towards Sgr A West. Absorption is also seen towards Sgr A* in the velocity ranges of +43 to -36 km s^{-1} and -142 to -151 km s^{-1}. The Northeast and Southwest torus components can be traced out to velocities of +139 and -151 km s^{-1}. A "Northwest" feature, starting northwest of the Western continuum arc, appears faintly at a velocity of +51 km s^{-1} and then migrates slightly southward until it merges with another feature at a velocity of +78 km s^{-1}. This other feature "Sweeps" in towards Sgr A* as the velocity decreases from +78 to +16 km s^{-1}.

INTRODUCTION AND OBSERVATIONS

Previous high resolution maps of OH towards Sgr A were obtained by the lunar occultation series in 1968 (Figure 1). They revealed that the continuum source is surrounded by a rotating and contracting cloud of dust and molecules (Sandqvist 1974)[1]. The infrared structure of this torus cloud was described in 1982 by Becklin, Gatley and Werner[2], while its kinematics was further discussed by Lester et al.[3] in 1981, Genzel et al.[4] in 1982 and Liszt, Burton and van der Hulst[5] in 1985. The "missing" Southeast molecular arm of the torus was detected and mapped with the rest of the torus in the J=3-2 HCO$^+$ line by Sandqvist, Wootten and Loren[6] in 1985.

We have studied the 18-cm OH distribution in the Galactic Center region near Sgr A in all four of the 1612, 1665, 1667 and 1720 MHz OH lines using the VLA in the A/B configuration. The observations were made on June 25 and 30, 1986 using 18 antennas and 64 frequency channels. The angular resolution in the 25' field of view is 4 arcsec, the channel frequency separation is 48.828 kHz resulting in a velocity resolution of about 9 km s^{-1} over a velocity range of about +/-280 km s^{-1}. The data was calibrated at the VLA using standard routines.

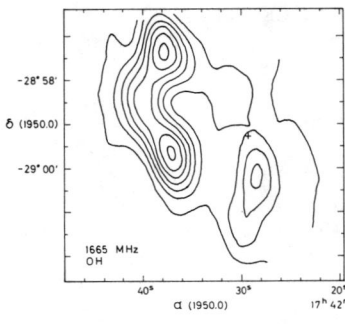 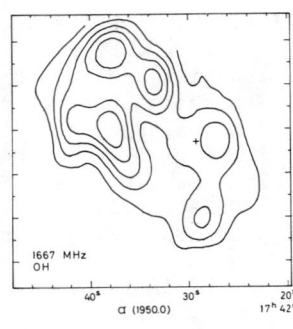

Figure 1. The velocity-integrated (-42 to +102 km s^{-1}) OH distribution (N_{OH}/T_{ex}). The cross marks the position of Sgr A*. (Sandqvist 1974)[1]

RESULTS AND DISCUSSION

Figure 2 shows three selected velocity channel maps (at +51, +25 and -1 km s^{-1}) covering the continuum region, which is also shown. The data has been hanned and the continuum map cleaned (Högbom 1974[7] and private communication). The outermost contour level in the spectral line aborption maps is -20 mJy/beam and the contour interval is -20 mJy/beam. For the continuum map, these values are 5 and 10 mJy/beam, respectively.

It is apparent from this figure that the gas in the +50 and +20 km s^{-1} molecular belt is seen clearly in absorption against the shell structure of Sgr A East but not against the spiral structure of Sgr A West. This may imply that the molecular belt lies between the two continuum components, behind Sgr A West and in front of Sgr A East.

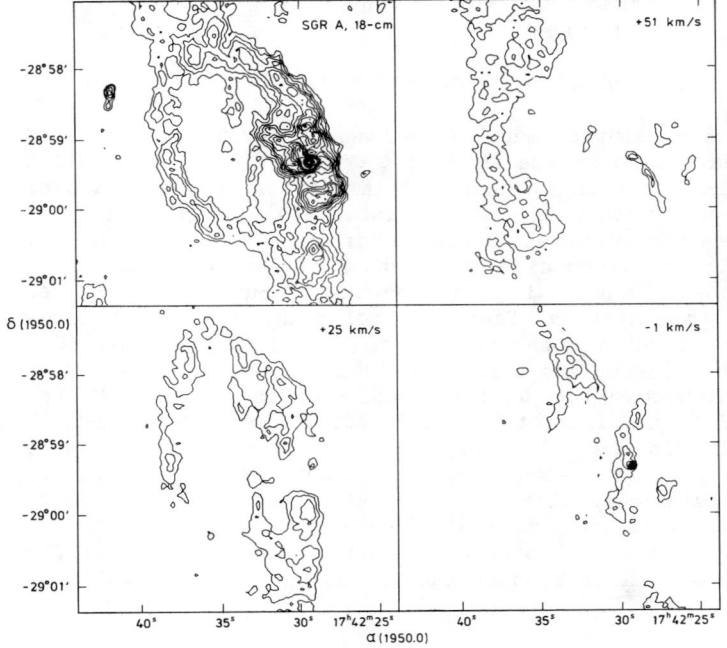

Figure 2. 1667 MHz OH spectral line absorption maps at +51, +25 and -1 km s^{-1}, and the 18-cm continuum map of Sgr A.

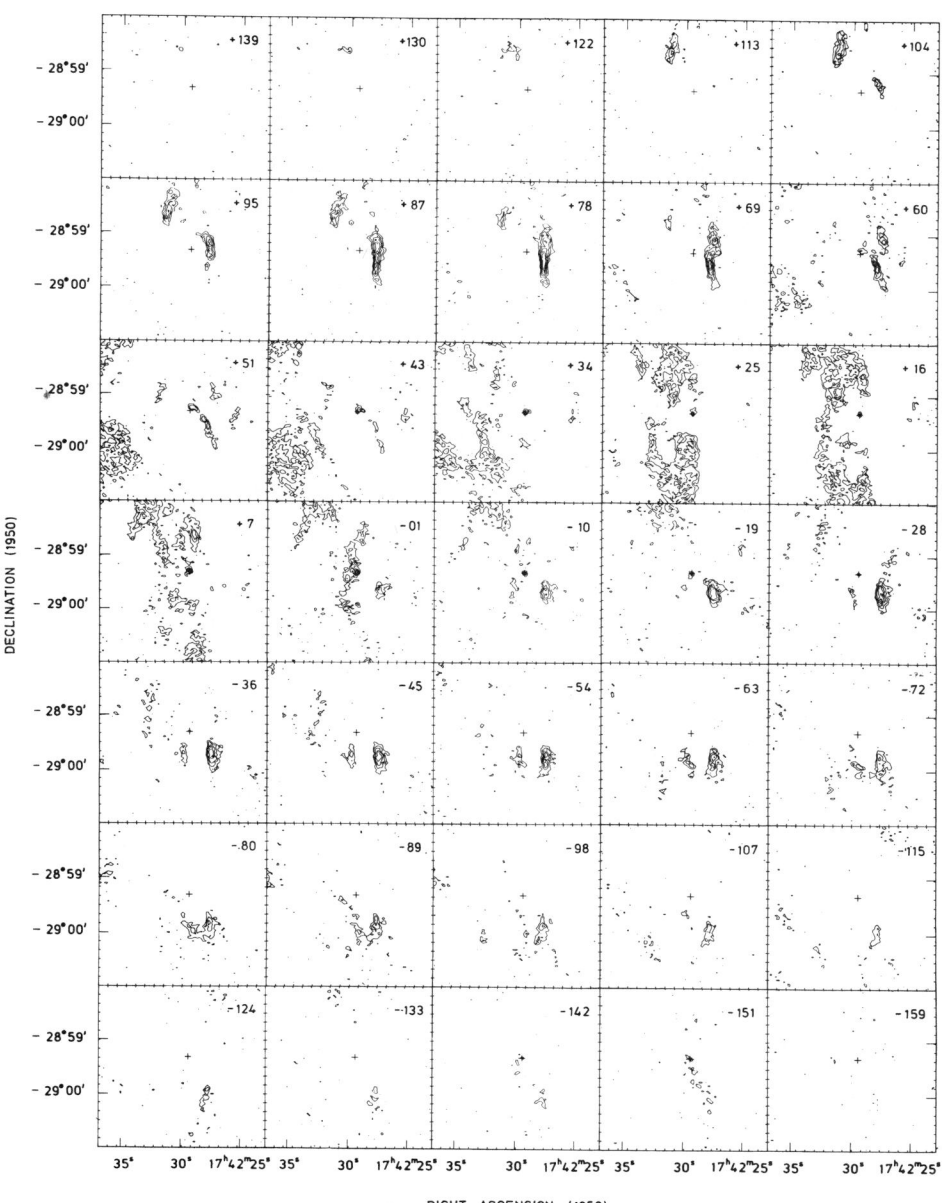

Figure 3. The 1667 MHz OH spectral line absorption maps. The radial velocity is given in the upper right hand corner. The cross marks the position of Sgr A*. The outermost contour level is -25 mJy/beam and the contour interval is -20 mJy/beam.

Yusef-Zadeh and Morris (1986)[8] have also presented evidence that Sgr A East lies behind Sgr A West. On the other hand, Whiteoak, Pankonin and Gardner (1983)[9] did find H_2CO absorption against Sgr A West at velocities of +41.5 to +49 km s^{-1}, so the picture is still not perfectly clear.

Figure 3 presents the 1667 MHz OH spectral line absorption maps of the nuclear torus region in the velocity range of +139 to -159 km s^{-1}. The outermost contour level is -25 mJy/beam and the contour interval is -20 mJy/beam. The angular resolution is 4 arcsec. A cross marks the position of Sgr A*; the radial velocity with respect to the local standard of rest is given in the upper right hand corner.

The Northeast and Southwest OH components, seen in the occultation maps, are very clear in the VLA maps. They can be traced out to velocities of +139 and -151 km s^{-1}, respectively. The signature of a rotating torus structure is especially clear in maps at velocities ranging from -63 to -98 km s^{-1}. Two new features, probably related to the torus but with distinct positive radial velocity components, can also be clearly identified. A "Northwest" feature, starting northwest of the Western continuum arc, appears faintly at a velocity of +51 km s^{-1} and then migrates slightly southwards until it merges with another feature at a velocity of +78 km s^{-1}. This other feature "Sweeps" in towards Sgr A*, in a most tantalizing manner, as the velocity decreases from +78 to +16 km s^{-1}. Finally, OH absorption is seen towards Sgr A* at all velocities in the ranges of +43 to -36 km s^{-1} and -142 to -151 km s^{-1}. Whiteoak, Gardner and Pankonin[9] found H_2CO absorption towards Sgr A* near 0 km s^{-1} but not between +40 and +50 km s^{-1}, whereas Liszt et al. (1983)[10] found HI in absorption towards Sgr A* at velocities between +40 and +60 km s^{-1}. The OH lines may give a more complete representation of the velocity extent of the gas absorption seen towards Sgr A*.

REFERENCES

1. Aa. Sandqvist, Astron. Astrophys. **33**, 413 (1974).
2. E. E. Becklin, I. Gatley and M. W. Werner, Astrophys. J. **258** 134 (1982).
3. D. F. Lester, M. W. Werner, J. W. V. Storey, D. M. Watson and C. H. Townes, Astrophys. J. (Letters) **248**, L109 (1981).
4. R. Genzel, D. M. Watson, C. H. Townes, D. F. Lester, H. L. Dinerstein, M. W. Werner and J. W. V. Storey, The Galactic Center, eds. G. Riegler and R. Blanford (American Institute of Physics, N. Y., 1982), p. 72.
5. H. S. Liszt, W. B. Burton and J. M. Van der Hulst, Astron. Astrophys. **142**, 237 (1985).
6. Aa. Sandqvist, A. Wootten and R.B. Loren, Astron. Astrophys. **152**, L25 (1985).
7. J. A. Högbom, Astron. Astrophys. Suppl. **15**, 417 (1974).
8. F. Yusef-Zadeh and M. Morris, Astrophys. J., Preprint (1986).
9. J. B. Whiteoak, F. F. Gardner and V. Pankonin, Mon. Not. Roy. Astron. Soc. **202**, 11P (1983).
10. H. S. Liszt, J. M. van der Hulst, W. B. Burton and M. Ondrechen, Astron. Astrophys. **126**, 341 (1983).

NH_3 in the Molecular Ring at the Galactic Center

James M. Jackson
Radio Astronomy Laboratory, Berkeley, Ca. 94720

Paul T. P. Ho
Harvard College Observatory, Cambridge, Ma. 02138

Alan H. Barrett
Massachusetts Institute of Technology, Cambridge, Ma. 02139

Abstract

Using the VLA, the NH_3 (J,K) = (3,3) inversion line has been mapped toward the Galactic center in a 2' field with 11" by 8" angular resolution ($\sim$ 0.5 pc at 10 kpc). A number of NH_3 condensations coincident with the molecular ring mapped in HCN by Gusten et al. (1986) are detected. These condensations are well localized (< 1 pc), have typical peak brightness temperatures $\sim$ 1 - 5 K, and have typical linewidths $\sim$ 15 km s^{-1}. We estimate the masses of the individual condensation to be on the order of 100 M$_\odot$. The differences in the distribution of HCN [1] and NH_3 may be caused by chemistry, by foreground absorption of the HCN due to intervening cold gas, or by differences in excitation. Detailed comparison with Ne II data [2] supports a model of photoionization of cloud material on the surfaces of the condensations facing the Galactic center. Assuming that the ionized gas flows toward the Galactic center and that the NH_3 emission defines the "mass-weighted" velocity, we conclude that the western side of the ring is behind the Galactic center while the eastern side is in front of the Galactic center.

Comparison with HCN Map

A map of the velocity-integrated NH_3 (3,3) emission is shown in Figure 1. The spectra refer to the labelled clumps, with the solid lines representing the NH_3 emission and the dashed lines the Ne II emission [2] at the same position. In Figure 2, the NH_3 map (stippled) is superposed on the HCN map synthesized using the Hat Creek millimeter wave interferometer [1]. There is fair overall agreement but distinct disagreement in detail. The condensations A, B, C, and G, which are strongest in the NH_3 emission, are either absent or fall on minima in the HCN emission. The remaining condensations are in good agreement with the HCN map. However, the bulk of the HCN emission to the north is not detected in NH_3. There are a number of possible explanations for this discrepancy: (1) Chemistry may be important, so that the relative abundance of various molecules varies from position to position. (2) Optical depth effects may be important. Condensations A, B, and G are at positions where the HCN spectra appear to suffer from absorption by intervening foreground gas. The poor correspondence to the HCN map may be due to "extinction" which is difficult to correct for in the HCN line. (3) Excitation effects may be important. The NH_3 (3,3) transition is more sensitive than the ground state HCN 1 - 0 transition to material at a higher temperature, probably material closer to the radiative excitation source. This would suggest that the HCN emission to the north, missing in the NH_3 map, may be further from the Galactic center than the apparent projected distance. The overall agreement in velocity between the HCN and NH_3 is quite good. The NH_3 emission appears clumpier than the HCN emission, perhaps due to differences in sensitivity.

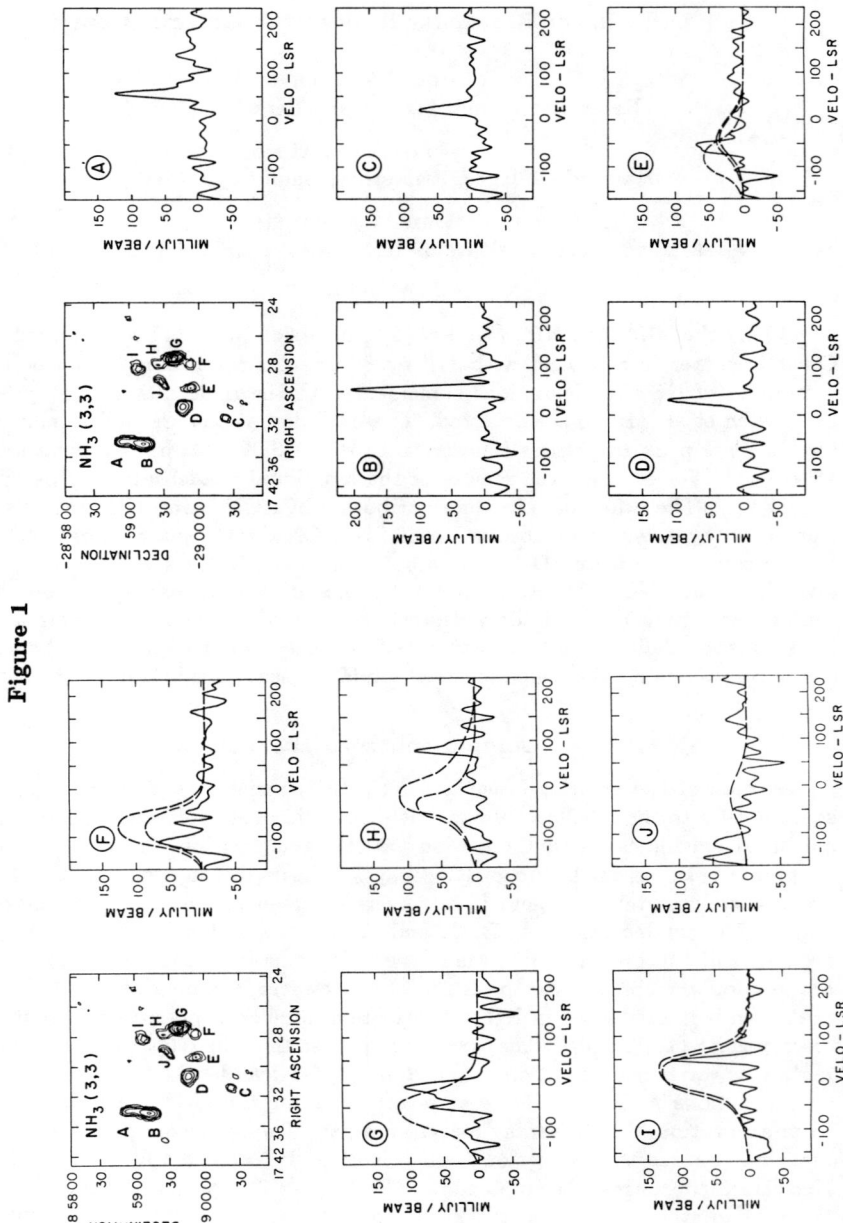

Figure 1

Figure 2

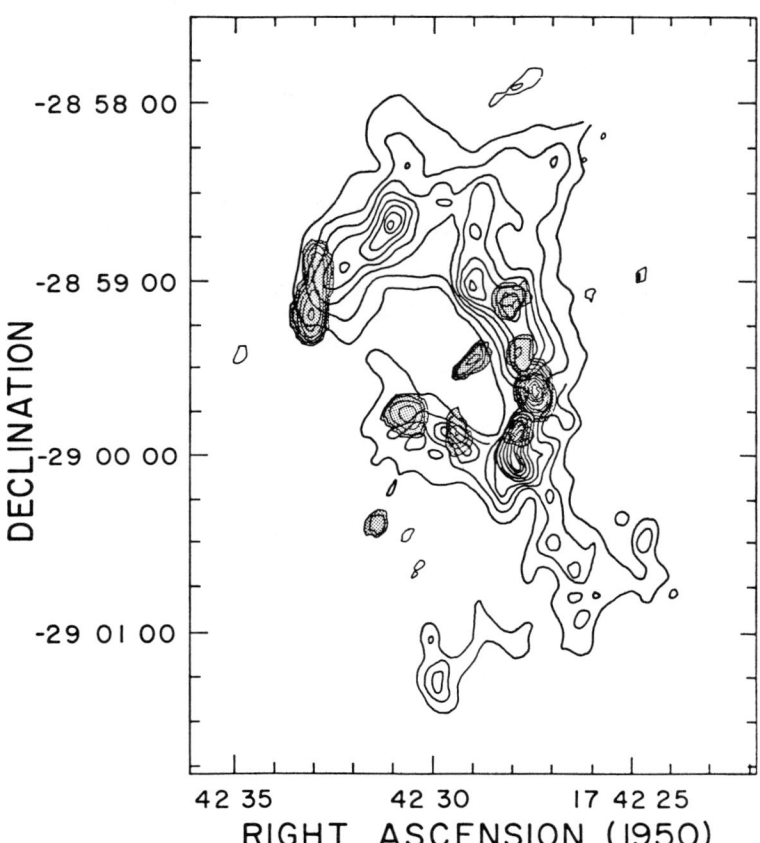

Comparison with the Ionized Gas

Figure 3 shows the NH_3 clumps superposed on the 1.3 cm continuum, which traces the ionized "spiral-like" features. The molecular gas seen in NH_3 and HCN lies on the boundary of the ionized gas. In every case the ionized material is on the side facing the Galactic center. Therefore, a central exciting source is strongly suggested. The Ne II spectra [2] in Figure 1 show that the ionized gas has much larger linewidths than the neutral gas. Further, on the western side the ionized gas appears consistently blueshifted with respect to the neutral gas. A possible configuration would be for the western condensations to be slightly behind the Galactic center. Then material stripped off from the molecular ring may fall toward the center with a net "blue" radial motion along the line of sight. The eastern condensations by a similar argument appears to be in front of the Galactic center. Material stripped off these condensations appear to move with a net "red" radial motion. The absence of neutral gas either in NH_3 or HCN associated with the northen arm suggests that this streamer may have completely moved away from the neutral material and is now very close to the Galactic center. A number of condensations (A, B, and C) are not associated with ionized gas. They may represent gas further removed from the Galactic center.

References

1. Gusten, R., Genzel, R., Wright, M.C.H., Jaffe, D.T., Stutzki, J., and Harris, A.I., preprint, (1986).

2. Serabyn, E. and Lacy, J.H., Ap. J. **293**, 445 (1985).

Figure 3

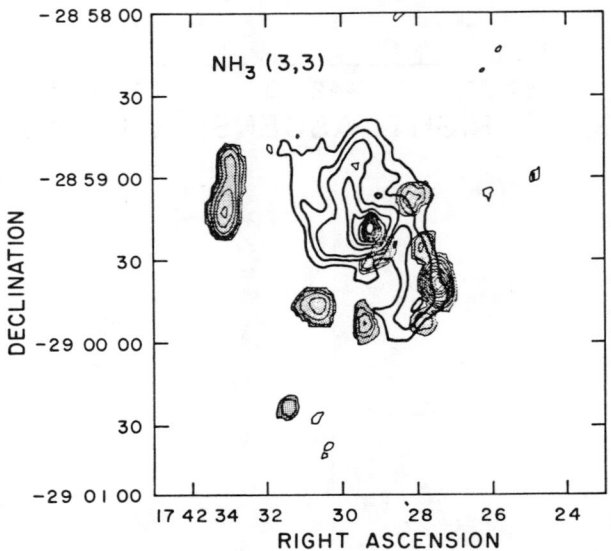

HAT CREEK APERTURE SYNTHESIS OBSERVATIONS OF THE CIRCUM-NUCLEAR RING
IN THE GALACTIC CENTER

R. Güsten, R. Genzel, M. C. H. Wright, D. T. Jaffe, J. Stutzki
and A. Harris   (U.C. Berkeley)

ABSTRACT

We have made aperture synthesis observations of HCN J = 1-0 emission and absorption in the central 5 pc of the Galaxy with 2" spatial and 4 km/s spectral resolution. The resulting maps show a clumpy ring of molecular gas which surrounds the ionized central 2 pc of the Galaxy. The molecular gas is dynamically coupled to the ionized gas in the central cavity ; the western arc of the radio continuum appears to be the inner surface of the molecular ring ionized there by UV radiation from the center, the northern arm may be ionised gas falling from the ring into the Galactic center. The molecular ring is warped with an inclination which changes from 50 to 75 degrees. The velocity field of the molecular gas in the inner 5 pc is that of strongly perturbed rotation. No overall radial motion of the ring greater than 20 km s^{-1} is apparent. There is a large local velocity dispersion (FWHM=40-50 km s^{-1} in a 4" beam) which indicates that the present features have a short dynamic lifetime ($<10^5$ y). The disk's turbulence may be maintained by dissipation of rotational energy and by energy input from the center. The neutral gas ring may be an accretion disk from which interstellar material falls into the central pc of the Galaxy.

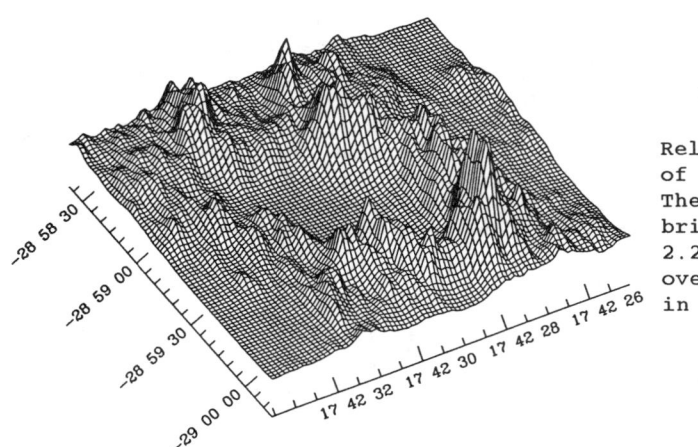

Figure 1.

Relief projection of HCN emission. The peak brightness is 2.2 K averaged over 300 km s^{-1} in a 2" beam.

OBSERVATIONS AND RESULTS

New data were obtained using the Hat Creek mm-interferometer and 512 channel digital correlator. We observed the HCN J=1-0 line at 88.632 GHz with 15 configurations of the three 6.1 m. antennas giving a 2" spatial and 4 km s^{-1} velocity resolution in a 3' field.

The velocity integrated HCN distribution (Figure 2) shows a complete ring of molecular gas which surrounds the ionized central 2 pc of the Galaxy. The major and minor axes are 4.8 x 2.5 pc, with the position angle of the major axis ~ 30°, i.e. approximately along the galactic plane. The inferred inclination is ~60°. The HCN ring has a sharp inner edge and can be traced out to 5 pc in the interferometer data. The ring is very clumpy with knots 5 - 10" in diameter and peak brightness temperatures 5 to 10 K. The HCN emission near the minor axis has deep self-absorption notches.

DISCUSSION

1 - comparision with radio continuum

The peak of molecular emission is located a few arcseconds outside the peak of ionized gas along the western arc. There is also a good correlation between individual knots of ionized and molecular gas. These results confirm the idea that the western arc is the inner, photoionized surface of a dense neutral ring exposed to the UV from the center[1,2,3]. Where the northern and bar intersect the molecular ring the HCN emission is supressed and the velocities inferred from the Ne II, 12.8μm line[2] for the northern arm agree well with the HCN velocities. These observations support the interpretation that the northern arm is a stream of material falling from the molecular ring towards the center[1,2]. The motions in the bar do not fit this model.

2 - comparison with other lines

The HCN ring is part of a more extended structure mapped in C^+ at 158 μm[4]. and in CO 1-0[5] The HCN, which requires higher densities for its excitation, is brightest near the inner edge; C^+ and CO 1-0 are sensitive to lower density material further out. A similar excitation gadient is found by comparing CO 7-6[6] and CO 1-0 emission. H_2, S(1) emission[7] shows the same overall distribution as the HCN, including the extra, redshifted cloud west of the center. This implies that the shock and/or UV excited regions emitting 2 μm H_2 are closely associated with the warm dense material.

3 - velocity distribution

The velocity field of the molecular gas in the inner 5 pc is that of strongly perturbed rotation[8]. There is no overall radial motion of the ring greater than 20 km/s, but the rotation is perturbed in several ways : 1) The gas has a large local velocity dispersion (FWHM=60 km/s). 2) The ring is warped with changes in position angle and inclination. 3) There is an extended , redshifted cloud in the western part of the ring which does not participate in the rotation. The high local velocity dispersion of individual clumps and the warp of the ring both imply that the present appearance of the ring has a lifetime < 10^5 years. The ring must have been recently formed, or mass, energy and clumpiness must be continuously replenished. The high dispersion and clumpiness may be maintained by slow contraction of the disk, feeding gravitational energy into turbulence.[8]

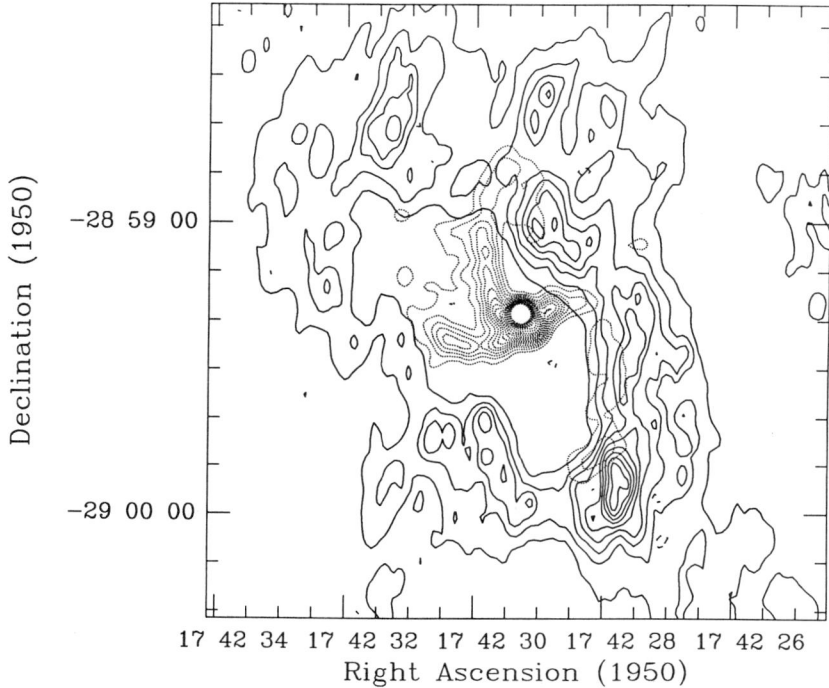

Figure 2. Superposition of the neutral ring mapped in HCN emission, and ionized gas mapped at 15 GHz^9. Solid contours of velocity-integrated HCN emission. The contour interval is 0.15 K averaged over 300 km s^{-1} in a 2" beam. Dotted contours of 15 GHz emission at 20 K intervals in a 3".6 x 3".4 beam.

REFERENCES

1. K. Y. Lo and M. J. Claussen, Nature, 306, 647. (1983)
2. E. Serabyn and J. H. Lacy, Ap. J. 293, 445. (1985)
3. R. Genzel, D. M. Watson, M. K. Crawford and C. H. Townes, Ap. J. 297, 766 (1985)
4. J. B. Lugten, R. Genzel, M. K. Crawford and C. H. Townes, Ap. J. 306, (1986)
5. E. Serabyn, R. Guesten, C. M. Walmsley, J. E. Wink and R. Zylka Astron. and Astrophys. submitted (1986)
6. A. I. Harris, D. T. Jaffe, M. Silber, and R. Genzel, Ap. J. (Letters) 294, L93. (1985)
7. I. Gatley, T. I. Jones, A. R. Hyland, R. Wade, T. R. Geballe and K. Krisciunas, Mon. Not. R. astr. Soc. 222, 299 (1986)
8. R. Guesten, R. Genzel, M. C. H. Wright, D. T. Jaffe, J. Stutzki and A. Harris, Ap. J. submitted (1986)
9. R. D. Ekers, J. H. van Gorkom, U. J. Schwartz, and W. M. Goss, Astron. and Astrophys., 122, 143. (1983)

ROTATING MOLECULAR RING AT THE GALACTIC CENTER

N. Kaifu, M. Hayashi, J. Inatani
Nobeyama Radio Observatory, Nobeyama, Minamimaki, Nagano, Japan

I. Gatley
National Optical Astronomy Observatory, Tucson, AZ85726-6732

ABSTRACT

The rotating molecular ring surrounding the Galactic center was found in the CS 2-1, HCN 1-0, HCO+ 1-0 and 13CO 1-0 lines. The 19" - 15" resolution maps of above 4 molecules show generally similar structure of the ring with the rotaion velocity of 100 km/sec and the radial velocity of roughly 50 km sec. The ionized gas feature fits very well to the inside edge of the molecular ring in its south part, but seems to sit in the central hole of the molecular ring in the north part. The structure and dynamics of the central 5 pc of the Galaxy are discussed.

OBSERVATIONS

The observations were made using the Nobeyama 45-m telescope in March 1985 (CS 2-1 line) and in June 1986 (HCN 1-0, HCO+ 1-0 and 13CO 1-0 lines). The observed area and the number of observed points are 1.5' x 3' and 144 points for CS, and 1.2' x 2.7' and 170 points for HCN, HCO+ and 13CO, respectively, with the grid spacing of 9". In the CS map the 18" grid was taken for outer region of the map. The beamwidths of the 45-m telescope are 17", 19", 19" and 15" for above four transitions, respectively.

The system noise temperature of the cooled mixer SSB receiver used for the CS 2-1 observations was 400 K typically, including the atmospheric effect. The HCN, HCO+ and 13CO maps were taken simultaneously using two cooled mixer receivers with perpendicular polarizations, which provided the SSB system temperature of 700 K and 500 K for the 90 GHz band and 110 GHz band, respectively. The 2000 channel wideband AOS with the frequency resolution of 250 KHz (corresponding to the velocity resolution of 0.8 - 0.7 km/sec) was used.

RESULTS AND DISCUSSION
(1. General Features)

A common ring structure was found in all these four lines, as can be seen in Figure 1. A central hole is evident especially in the HCN, HCO+ and CS maps, corresponding to the position of the nucleus, thus only poor amount of cold gas seems to exist in the central 1 pc of the Galactic center. The ring is elongated to the direction of the Galactic plane with the major/minor axis ratio of 2:1.

The size of the ring (5 pc) is larger than that of the central ionized gas feature, and the ionized gas seems to sit in the central hole. The ring can be seen most clearly in the HCN map (Figure 1-A). However the south-east side of the ring is considerablly weak in the HCN and HCO+ maps while it is not clear in the CS and 13CO. The 13CO map shows the most poor structure.

The observed HCN ring generally coincides to the HCN map taken with the Hat Creek interferometer(1) except that the 45-m map shows rich extended components ovelapping to the sharp ring structure. Also Sandquvist et al.(2) observed the HCO+ ring feature with the 43" beam. The shocked H_2 gas observed by Gatley et al.(3) shows very similar ring-like distribution, though the east-south part is lacking.

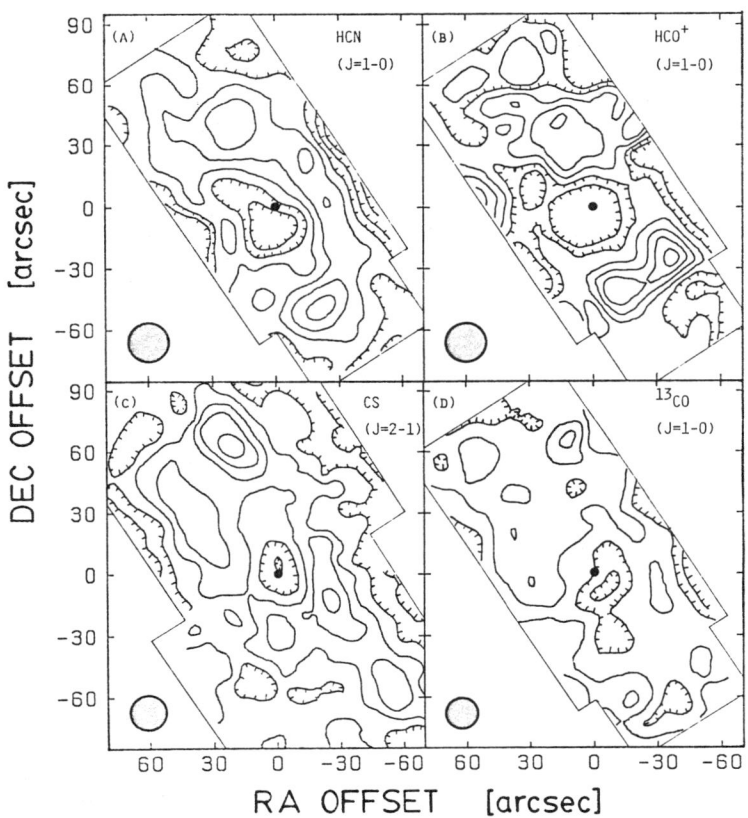

Figure 1. Total integrated maps of Sgr A in the HCN 1-0, HCO+ 1-0, CS 2-1 and 13CO 1-0 transitions, with the contour intervals of 30, 20, 10 and 20 K km/sec, respectively. Observed area and HPBW are shown together. The central position (small filled circle) is 17h42m29.29s, -28°59'17.6" (1950).

(2. Kinematics)

The velocity-divided HCN map (Figure 2) show the clear rotation of the ring. The rotation velocity of about 100 km/sec coinsides well to the values obtained from the previous observations (CO 1-0 (4), CO 7-6(5), CII 158 m(6), H_2 2 m(7), HCN(1) and HCO+(2). The rotaion velocity does not require massive unvisible object in the center, if we adopt the stellar mass of 4×10^6 Mo within the radius of 1 pc which was derived by Oort (8). Furthermore, as the ring seems to move radially as described in the following part, the rotation velocity can not be used for the estimation of the central mass. Note that the north-west peak seen in the 60-90 km/sec map in Figure 2 may not be a part of the ring.

The radial motion of about 50 km/sec is seen most clearly in the CS map, and is seen in the HCN and HCO+ maps too. It is hard to determine the aspect of the radial motion, however the expanding motion may fit to the following observational evidences: (1) A distinct dip is seen in the south-east part of the ring of HCN and HCO+, which shows the negative velocity (see Figures 2 and 3). As the 1-0 transitions of HCN and HCO+ are easily self-absorbed by the cold gas we can assume that the dips are due to the absorption by the cold outer part of the ring lies in this side of the center. (2) The 2 m H_2 emission is observed in the north-west side of the molecular ring (slightly inner part of the ring), but not in the south-east side, while the 63 m OI emission is seen in the inner part of the south-east side(9). These facts may be explained by the absorption of the H_2 emission due to the outer cold molecular ring. Therefore we suggest the expansion motion of the ring, though the possibility of the contractiong motion described by several authors (e.g. Gatley et al.(3) is not ruled out yet.

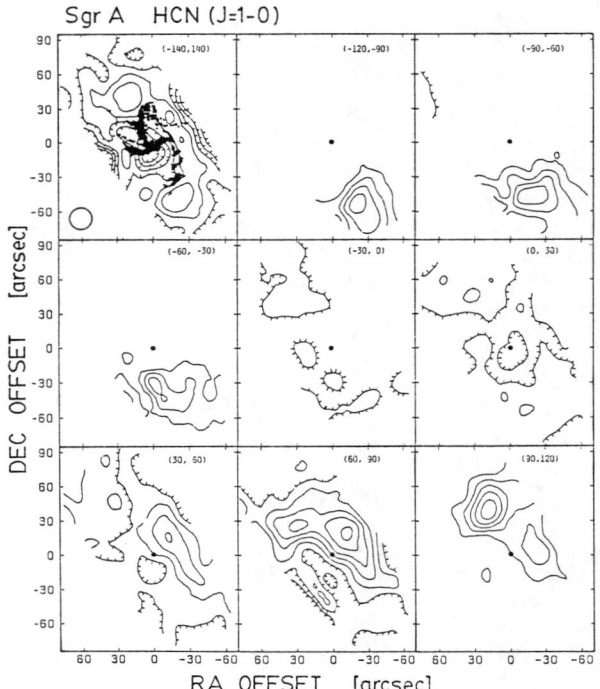

Figure 2. HCN velocity maps of Sgr A. The integrated velocity range is shown in each velocity map. The contour interval is 0.3 K km/sec except for the total integrated map in the top-left, in which the ionized gas features[9] is shown together.

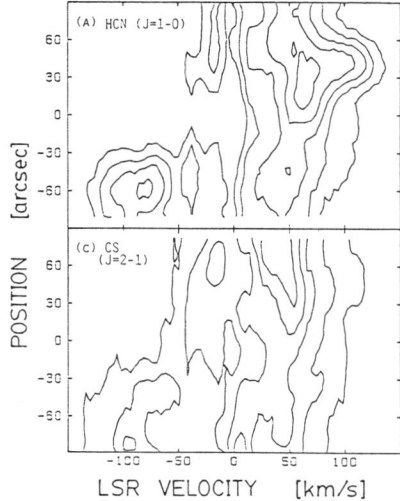
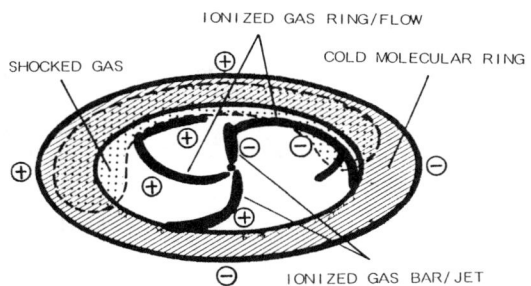

Figure 3 (left). Position-velocity maps of HCN and CS in the direction of major axis.

Figure 4 (top). A sketch of central 5 pc of the Galaxy. The signs in circles denote the direction of Vr of cold and ionized gas features.

(3. Structure of the Central 5 pc)

Figure 4 shows a sketch of the central 5 pc of the Galaxy. The ionized gas and warm/hot dust seem to occupy the inside of the molecular ring. The ionized gas feature fits very well to the inner edge of the molecular ring in its south-west part. The shocked H_2 gas found in this region(3)(7) also show the similar distribution, and the velocities of cold molecules, ionized gas and shocked gas are roughly the same. Thus in the south-west part it is very plausible that the hot gas actually hits to the inner edge of the cold molecular ring and causes the observed shock.

However, in the northern part the structure is less clear. The ionized features are complicated and seem to fill the hole of the molecular ring. The velocities of features in this region are all positive and are probably caused mainly by the rotation, but some part of the ionized features in this region may be originated by different mechanism to that of the southern ionized gas feature.

The central "bar" like feature of the ionized gas and IR emission shows very high opposite-direction velocity with respect to the molecular ring. If the molecular ring is expanding as mentioned in the former section the central "bar" can be interpreted as the jet being ejected from the nucleus region.

(REFERENCES)
1. R. Gusten et al.(1986), preprint
2. Aa. Sandqvist et al., Astron. and Astrophys. 152, L25 (1985)
3. I. Gatley et al., Mon.Not.R.Astr.Soc. 222, 299 (1986)
4. H. S. Liszt et al., Astron. and Astrophys. 142, 237 (1985)
5. A. I. Harris et al., Astrophys. J.(Letters) 294, L93 (1985)
6. J. B. Lugten et al., Astrophys. J. 306, 691 (1986)
7. I. Gatley et al., Mon. Not. R. Astr. Soc. 210, 565 (1984)
8. J. H. Oort, Ann. Rev. Astron. Astrophys. 15, 295 (1977)
9. R. Genzel et al., Astrophys. J. 297, 766 (1985)
10. K. Lo and M. J. Claussen, Nature 306, 647 (1983)

THE CO (J=2-1) DISTRIBUTION IN THE INNER 10PC OF THE GALAXY

Y. Fukui
Nagoya University, Nagoya 464, Japan

E. Churchwell
University of Wisconsin, Madison, WI 53706

ABSTRACT

We report observational results of the J=2-1 CO emission in the 5' x 5' region toward Sgr A*. The fully sampled high angular resolution (30") data reveal considerable details of the neutral gas "disk" in the central 10pc of the Galaxy. Although the CO distribution is generally consistent with a tilted rotating disk, the present data indicate that the distribution is significantly asymmetric with respect to Sgr A*. The CO rotation curve derived from the negative velocity CO lobe indicates that the mass in the inner 10pc increases linearly with radius from 3pc to 10pc.

INTRODUCTION

Molecular line emission localized in the central 10pc of the Galaxy was detected in the 3-millimeter HNC emission[1]. The purpose of the present study was to investigate details of this molecular gas with higher angular resolution of 30"(=1.5pc at the distance of 10kpc). To achive this goal we used the J=2-1 CO and ^{13}CO emission at 1.3 millimeters.

OBSERVATIONS

Observations were made by using the NRAO* 12m telescope on Kitt Peak in June 1984. The main beam size of the telescope was 33" at 1.3 millimeters, and observations of the CO emission were made with a grid spacing of 20" over a 5'x5' area centered on Sgr A*. The ^{13}CO emission was also mapped in the inner 2'x1' area with the same grid spacing.

RESULTS

1. DISTRIBUTION

The first panel of Fig. 1 shows the CO distribution in the v_{LSR} range of -120 to -100 km/s and 100 to 120km/s. Two dominant features delineate double lobes tilted to the galactic plane by about 20 degrees. The velocity distribution is consistent with rotation as suggested by previous studies[3,4,7].

*The National Radio Astronomy Observatory is operated by Associated Universities Inc., under contact with the National Science Foundation.

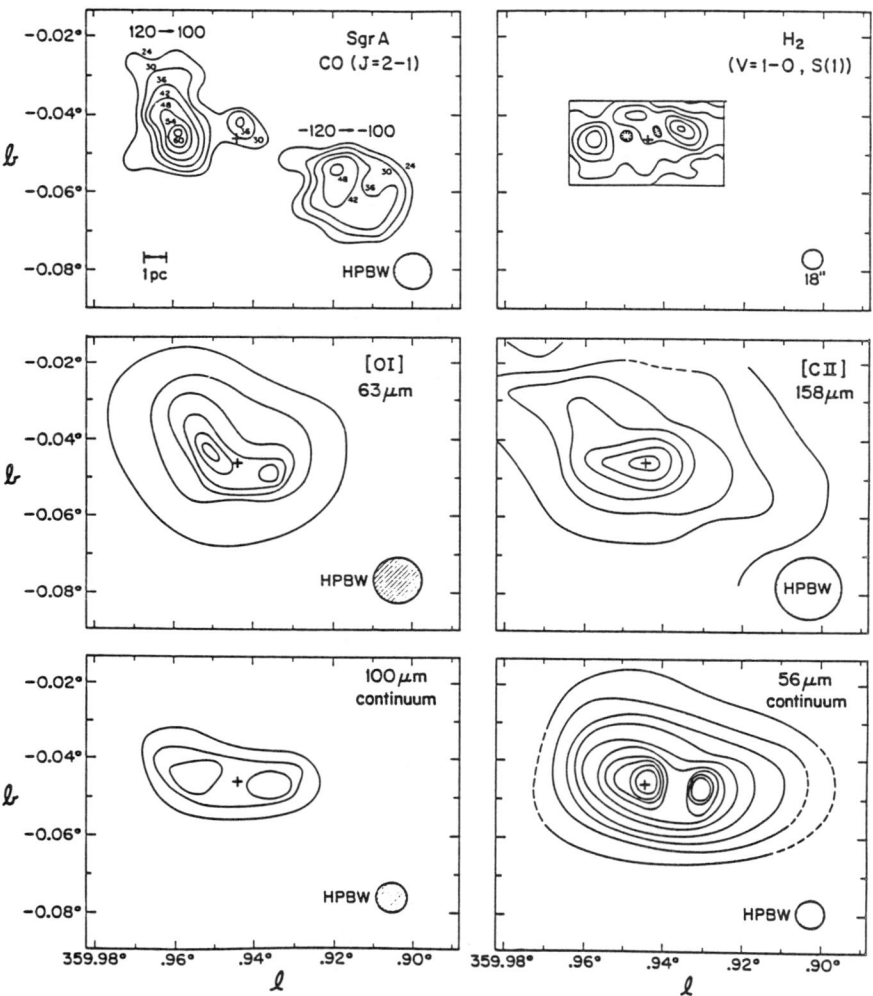

Figure 1. The CO (J=2-1) distribution of Sgr A is compared with other maps; shocked molecular hydrogen[2], neutral atomic oxygen[3], ionized atomic carbon[4], and far infrared emission[5,6]. Contour unit of the CO map is K km/s.

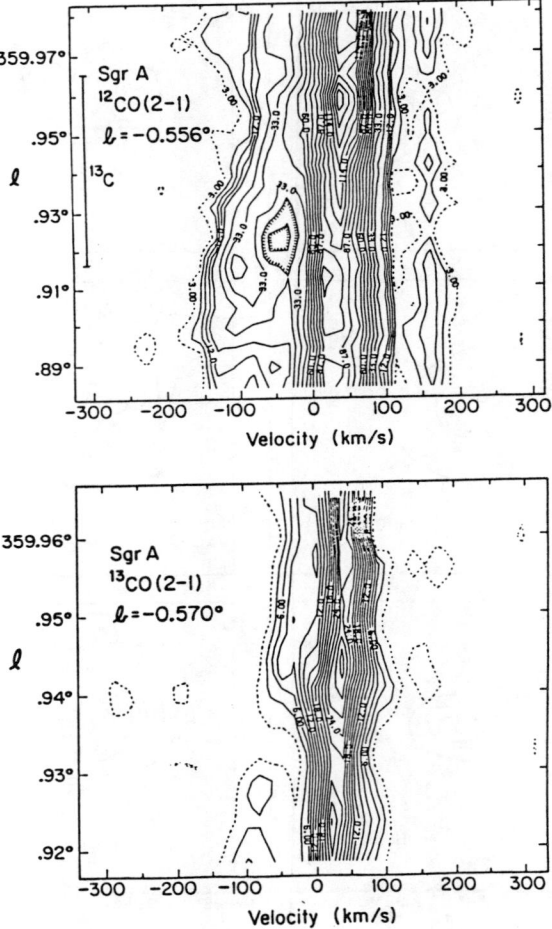

Figure 2. Velocity vs. galactic longitude map of CO (J=2-1). Contour unit is Kelvin in corrected antenna temperature.

The CO map reveals details not apparent in previous studies; (i) the distribution is asymmetric with respect to Sgr A*, that is, the separation of the negative lobe from Sgr A* is twice as large as that of the positive lobe, and (ii) there is a third small peak 20" north of Sgr A*. The asymmetry is not due to pointing error.

In the other panels of Fig. 1 we present distribution of shocked molecular hydrogen[2], warm atomic gas[3,4], and dust[5,6]. The peak of the positive lobe is coincident with one of the molecular hydrogen peaks, and the third CO peak is located in the northern part of the molecular hydrogen ring. The tilt indicated by the atomic carbon is very similar to that of the CO lobes, and the primary peak of the neutral oxygen is located just inside the positive CO lobe. These correspondences suggest that the atomic and molecular gas are physically associated with each other.

2. Kinematics

Fig. 2 shows a velocity vs. galactic longitude map taken at b=-0°.556. The low velocity emission is dominant at v_{LSR}= 0-100km/s, making it impossible to see the positive CO lobe at v_{LSR} 100km/s. The negative lobe, on the other hand, appears clearly at l=359°.89-359°.94. The peak CO velocity is almost uniform, -100km/s, at l=359°.89-359°.92.

3. Mass Distribution

We use the CO velocity curve to derive mass distribution by assuming that the velocity represents equilibrium rotation. For an inclination angle of 65° of the disk, we estimate that the total mass increases linearly from $8 \times 10^6 M$. at r=3pc to $2.5 \times 10^7 M$. at r=10pc. This distribution is less steep than suggested on the basis of the far-infrared data[4,7], weakening evidence for a central massive object.

A detailed account of this work will be presented elsewhere.
This research was financially supported by the Grant-in-Aid of the Ministry of Education, Science, and Culture of Japan (No.57420003).

REFERENCES

1. Y. Fukui, H. Ogawa, and S. Deguchi, Ap.J. **275**, L55 (1983).
2. I. Gatley, T. J. Jones, A. R. Hyland, R. Wade, T. R. Geballe, and K. Krisciunas, Mon. Not. R. astr. Soc. **222**, 299 (1986).
3. R. Genzel, D. M. Watson, M. K. Grawford, and C. H. Townes, Ap.J. **297**, 766 (1985).
4. J. B. Lugten, R. Genzel, M. K. Crawford, and C. H. Townes, Ap.J. **306**, 691 (1986).
5. E. E. Becklin, I. Gatley, and M. W. Werner, Ap.J. **258**, 135 (1982).
6. G. H. Rieke, C. M. Telesco, and D. A. Harper, Ap.J. **220**, 556 (1978).
7. A. I. Harris, D. T. Jaffe, M. Silber, and R. Genzel, Ap.J. **294**, L93 (1985).

CS EMISSION FROM THE GALACTIC CENTER RING

Neal J. Evans II
Department of Astronomy and Electrical Engineering Research Laboratory
The University of Texas at Austin

The ring of neutral material which surrounds the ionized central 2 pc of the Galaxy has been observed in a number of spectral lines, including OI (cf. Genzel et al.)[1]; CO, J = 1 → 0 (Liszt et al.)[2]; CO, J = 7 → 6 (Harris et al.)[3]; NH_3 (Serabyn and Güsten)[4], and HCN, J = 1 → 0 (Güsten et al.)[5]. From the J = 7 → 6 CO data, Harris et al.[3] argued that the material is dense (n ~ 10^5 cm^{-3}) and warm (T ~ 150 - 450K) and that the pressure decreases with distance from the center. Recently, Serabyn et al.[6] reported the detection of CS J = 2 → 1 emission from the ring.

Because CS has proven to be a good probe of density (Snell et al.[7]; Mundy et al.)[8], observations of higher J transitions of CS were made with the MWO* 5 m telescope. The beam was ~ 55" and the forward spillover and scattering efficiency $\eta_{fss} = 0.90$.

As shown in Figure 1, the J = 5 → 4 line was indeed detected. Observations at several points along the galactic plane showed the expected shift in velocity if the emission is due to the ring. An attractive feature of this transition is that the clouds at 20 and 50 km s^{-1} which produce confusing emission in the CS J = 2 → 1 and CO J = 1 → 0 and 2 → 1 lines do not seem to emit strongly in the J = 5 → 4 CS line. Moreover, the narrow foreground absorption features which appear in the HCN J = 1 → 0 map (Gusten et al)[5] do not appear in the spectra of CS J = 5 → 4. Thus, a high resolution map in this line should reveal the ring kinematics relatively clearly.

The current observations can be combined, in a preliminary way, with the 20" beam CS J = 2 → 1 data of Serabyn et al.[6] to provide a rough estimate of the density. Because the beam sizes are different, the ratio (R) of J = 2 → 1/J = 5 → 4 emission is not yet well determined. If we simply take the ratio of the observed T^*_R's, we obtain R = 2.8. Since there is structure smaller than the 55" beam used here, we almost certainly have R < 2.8. If the J = 2 → 1 line is not beam-diluted in the map of Serabyn et al.[6], but is diluted by a factor of 2 (a crude estimate from the 2 → 1 maps) in the 5 → 4 line, we have 1.4 ≲ R < 2.8.

----

* The Millimeter Wave Observatory is operated by the Electrical Engineering Research Laboratory of the University of Texas at Austin, with partial support from the National Science Foundation and McDonald Observatory.

Excitation and radiative transport models (Figure 2) were constructed in the LVG approximation and indicate that this range of line ratios, together with the strength of the CS 2 → 1 line, implies densities slightly exceeding 10^5 cm^{-3}, consistent with earlier estimates (Harris et al.[3]). If the filling factor for the J = 2 → 1 emission is not too small, the emission is optically thin, and the inferred column density of CS per unit velocity is about 5×10^{12} cm^{-2}/km s^{-1}. Because this line ratio is a strong function of density, but a weak function of kinetic temperature and CS abundance, in the regime of interest, further studies with matched beams of these lines should be able to establish the density of the neutral ring more securely and to separate the effects of density and temperature gradients in producing the pressure drop with radius inferred by Harris et al.[3].

REFERENCES

1. R. Genzel, D.M. Watson, M.K. Crawford and C.H. Townes, Ap. J., 297, 766 (1985).
2. H.S. Liszt, J.M. van de Hulst, W.B. Burton and M.P. Ondrechen, Astr. Ap., 126, 341 (1983).
3. A.I. Harris, D.T. Jaffe, M. Silber and Genzel, R., Ap. J. (Letters), 294, L93 (1985).
4. E. Serabyn and R. Güsten, Astr. Ap., in press (1986).
5. R. Güsten, R. Genzel, M.C.H. Wright, D.T. Jaffe, J. Stutzki and A.I. Harris, preprint (1986).
6. E. Serabyn, R. Güsten, C.M. Walmsley, J.E. Wink and R. Zylka, Astr. Ap., in press (1986).
7. R.L. Snell, L.G. Mundy, P.F. Goldsmith, N.J. Evans II and N.R. Erickson, Ap. J., 276, 625 (1984).
8. L.G. Mundy, R.L. Snell, N.J. Evans II, P.F. Goldsmith and J. Bally, Ap. J., 306, 670.(1986).

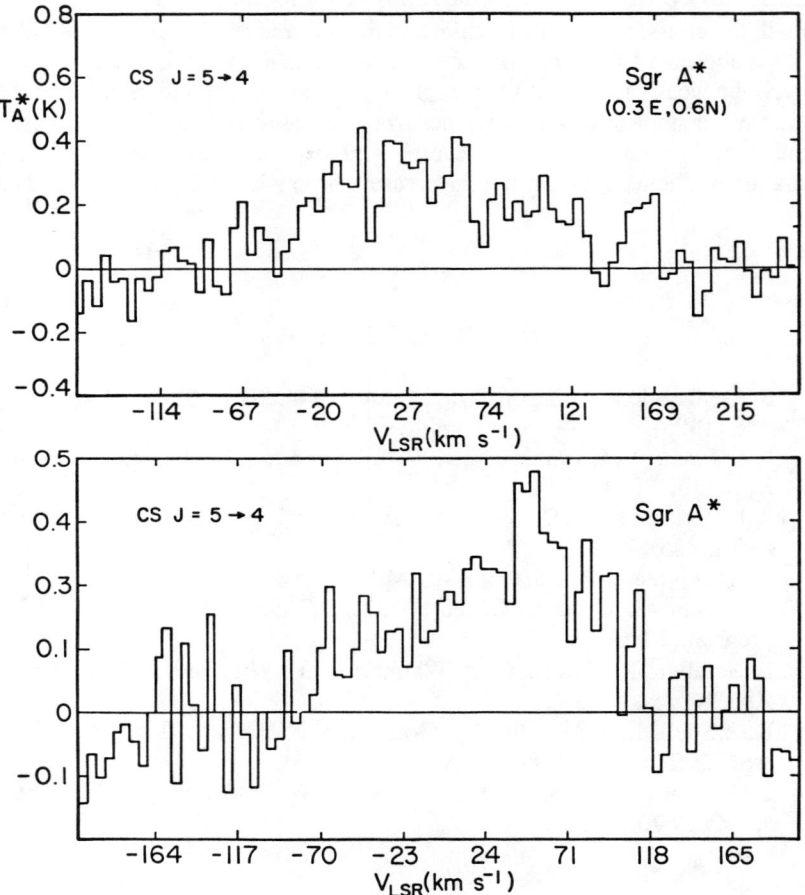

Figure 1: Spectra of the CS J = 5 → 4 line are presented for a position on the compact radio source and at a position with $(\alpha, \delta) = (0.'3 \, E, 0.'6 \, N)$. The data were taken with 1 MHz resolution, but four adjacent channels were averaged to produce the spectra shown here. The data were not smoothed. Note that the velocity and temperature scales are not the same for the two plots.

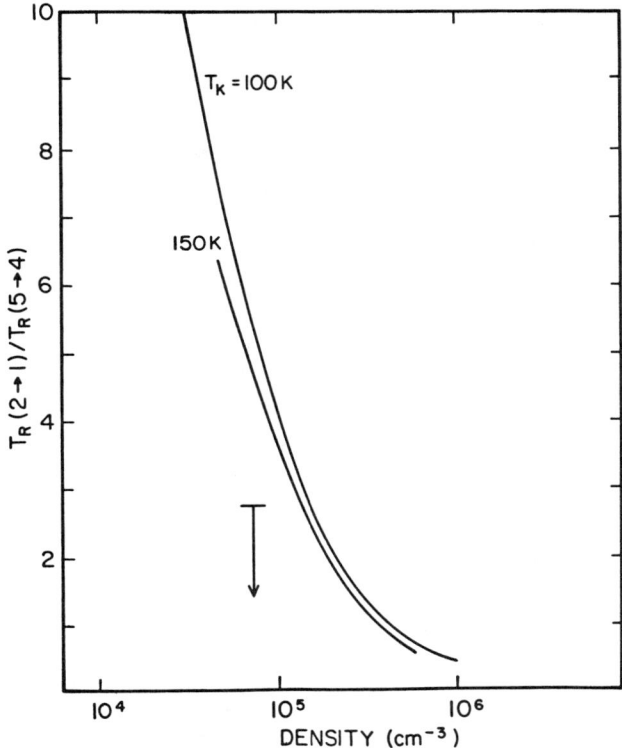

Figure 2: The ratio of the radiation temperatures of the $J = 2 \to 1$ and $J = 5 \to 4$ transitions is plotted versus the log of the density for two values of kinetic temperature. The arrow at the lower left indicates the range of observational ratios, $1.4 \lesssim R < 2.8$.

EXCITATION GRADIENT OF THE MOLECULAR GAS IN THE SGR A CIRCUM-NUCLEAR RING

J.B. Lugten, G.J. Stacey, A.I. Harris*,
R. Genzel*, and C.H. Townes
University of California, Berkeley, CA 94720

ABSTRACT

We present new measurements of CO submillimeter (J = 7 → 6) and far-infrared (J = 14 → 13, J = 21 → 20) lines in the galactic center. The new data, together with earlier CO measurements, are used to derive the physical conditions of the dense, warm molecular gas at different positions in the Sgr A circum-nuclear ring. From the inner edge of the ring at R ~ 1.7 pc to R ~ 6 pc the gas pressure decreases proportional to $R^{-1.6 \pm 0.4}$. The best fit excitation curves indicate that both density and temperature decrease with distance from the center. There is also an azimuthal excitation gradient in the ring. If the molecular gas is directly heated from the center by UV radiation or mass outflow the decrease of excitation is a natural consequence of the R^{-2} dependence of the energy density. In the more likely scenario of shock heating fed by the large turbulence of the gas throughout the ring, the excitation gradient may reflect variations in gas density and heating efficiency.

INTRODUCTION

Observations of line[1] and continuum[2] emission show that a ring of gas and dust surrounds the galactic center. The ring or disk is thin, extends to a radius of R ~ 8 pc or more, and surrounds a central cavity of R ~ 1.7 pc which, in comparison, is quite empty of gas and dust. The dominant motion of gas in the disk is rotation about the center although small scale turbulent motions are significant. Gas in the ring has been traced in a large number of emission lines. The inner edge of the ring contains ionized gas[3]; most of the ring is predominantly neutral gas. The gas is warm and dense, with substantial variation in excitation from one position to another.

OBSERVATIONS

The CO far-infrared lines (J ≥ 13) were observed using the U.C. Berkeley tandem Fabry-Perot spectrometer[4]. Observations were made at an altitude of 41,000 feet from the Kuiper Airborne Observatory on flights from Christchurch, New Zealand. The J = 14 → 13 line at 186 µm was observed on April 29, 1986, with a 65" beam. The J = 21 → 20

* now at the Max-Planck-Institut fur Physik und Astrophysik, D-8046 Garching bei Munchen, Federal Republic of Germany

line at 124 μm was observed on May 3, 1986 with a 55" beam. The CO J = 7 → 6 line at 370 μm was observed using the U.C. Berkeley Physics Department heterodyne spectrometer[5] on the IRTF and UKIRT telescopes on Mauna Kea, Hawaii on May 22, 1986 and June 17 and 19, 1986. The beam sizes were approximately 32" on the IRTF and 25" on the UKIRT. A total of 4 positions were observed as shown in Figure 1.

CONCLUSIONS

The main results are as follows:
1) At the inner edge of the molecular ring (R ~ 1.7) the CO far-infrared and submillimeter line emission can be well fitted by a single component model with gas temperature ~400 K and hydrogen density 5×10^4 cm^{-3} (Figure 2). The marginal detection (~3σ) of the 21 → 20 line provides good upper limit to the 21 → 20 line strength and suggests that there may also be a component of hotter (T ≥ 700 K) CO gas. The volume filling factor of the warm gas is ~0.1, indicating a clumpy structure.
2) The excitation of the gas decreases with radius from the center. This is shown in Figures 2 and 3. The gas pressure

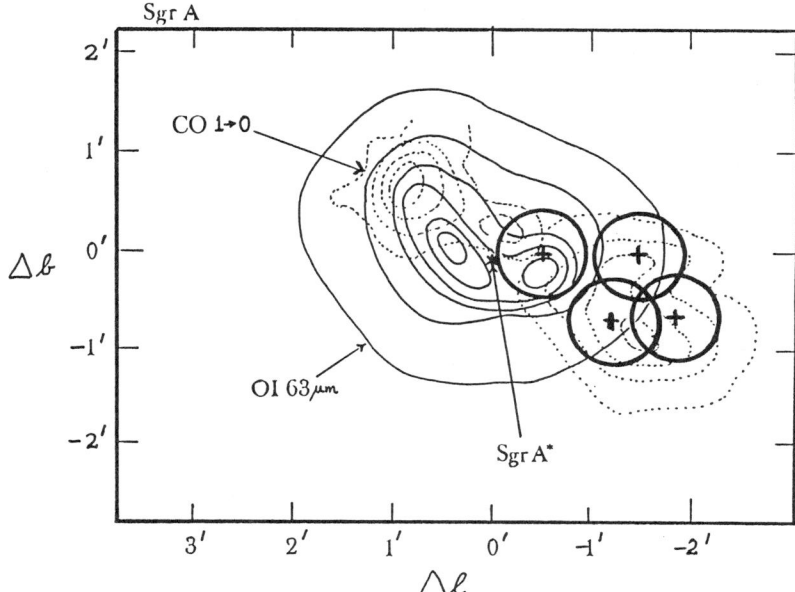

Figure 1. Positions in the galactic center where far-infrared and submillimeter CO lines were observed (heavy circles), superposed on the integrated 63 μm OI map[1] (light contours) and the CO 1 → 0 map[7] at 80 ≤ $|v_{LSR}|$ ≤ 120 km s^{-1} (dotted contours).

decreases by a factor of 8 ± 2 between 1.7 and 6 pc radius (pressure ∝ $R^{-1.6 \pm 0.4}$). The best fit excitation curves (Figure 3) indicate that both temperature and density decrease with distance from the center. The volume filling factor, however, appears to increase with radius.

3) There is an azimuthal excitation gradient (Figure 3). At R ~ 4.5 pc, gas near the galactic plane (Δl = -90", Δb = 0) has twice as

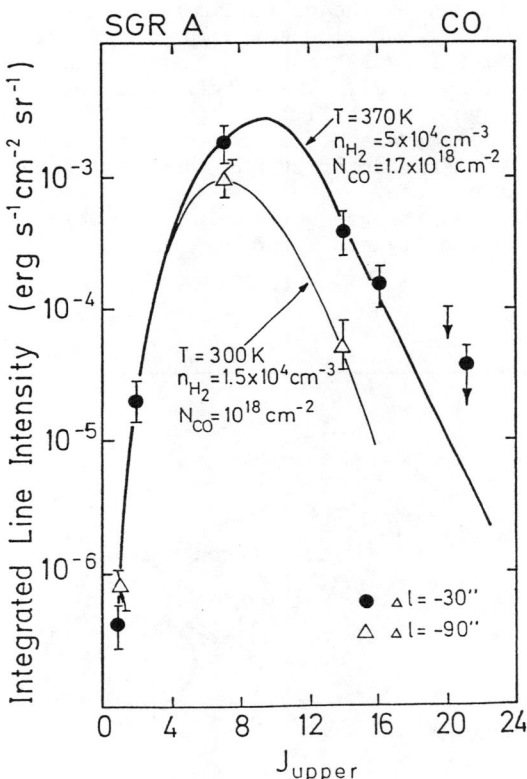

Figure 2. The distribution of observed CO integrated line intensities as a function of J at two positions in the ring. The data taken at a longitude offset, Δl = -30", from SgrA*/IRS 16 are indicated by black dots, while those at Δl = -90" are indicated by open triangles. The data from this paper (the 14 → 13 and 21 → 20 lines) are combined with several other measurements (the 16 → 15 line[1], the 7 → 6 line[8] and 21 → 20 line[8], the 2 → 1 line[9] and the 1 → 0 line[7].) Best fit LVG models are indicated for the two data sets. Note that the 21 → 20 (124 μm) line at Δl = -30" is only marginally detected at ~3σ significance.

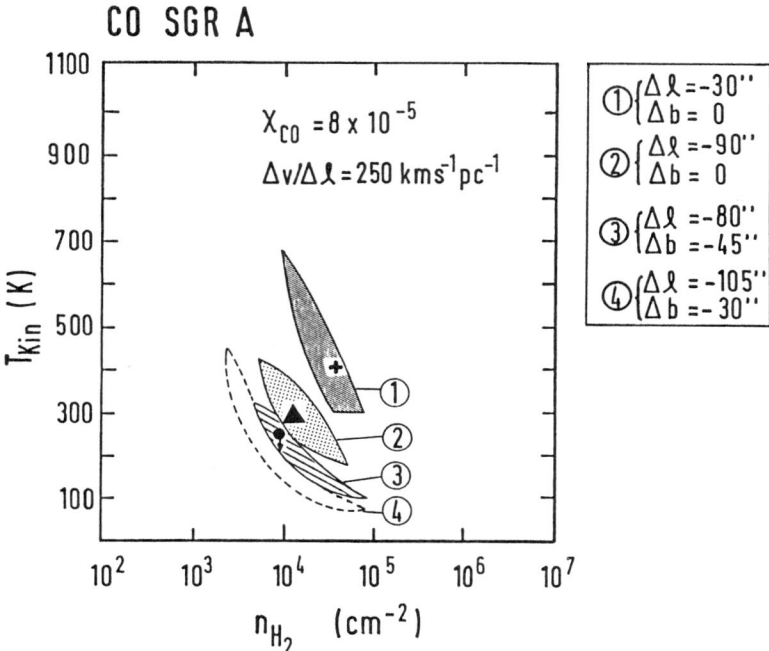

Figure 3. The density of molecular hydrogen and the gas temperature derived from the CO data at different positions in the disk. The LVG models assume a velocity gradient of 250 km s^{-1} pc^{-1} and a CO/H$_2$ abundance of 8×10^{-5}. The data clearly show an excitation gradient between 1.7 and 6 pc from the center.

much pressure as gas near the kinematic major axis (Δl = -75", Δb = -45").

4) The presence of warm CO gas throughout the circum-nuclear ring, the large local line widths Δv ~ 80 kms^{-1} in a 25" to 30" beam) and the radial falloff of pressure are well fit by a scenario[1,6] where the CO emission originates in slow cloud-cloud shocks driven by the large turbulence of the ring. The turbulence may be fed by dissipation of the strong differential rotation of the ring. The azimuthal gradient may reflect large scale density variations, or variations in heating efficiency. Alternatively, but perhaps less likely, the molecular gas may be excited by direct UV- photo heating or by mass outflow from the center.

We thank D. Jaffe, J. Stutzki, R. Gusten, H. Rothermel and the KAO, IRTF, and UKIRT staff for their assistance. This work was supported by NASA grant NAG 2-203 and NSF grant AST83-51381 to R. G.

REFERENCES

1. R. Genzel, D.M. Watson, M.K. Crawford, and C.H. Townes 1985, Ap. J. <u>297</u>, 766.

2. E.E. Becklin, I. Gatley, and M.W Werner 1982, Ap. J., <u>258</u>, 135.

3. E. Serabyn and J.H. Lacy 1985, Ap. J., <u>293</u>, 445.

4. J.B. Lugten, M.K. Crawford, and R. Genzel 1987, in prep.

5. A.I. Harris, D.T. Jaffe, and R. Genzel 1986, Proceedings of S.P.I.E. meeting held in Cannes, France, Dec. 1985.

6. R. Gusten, R. Genzel, M.C.H. Wright, D.T. Jaffe, J. Stutzki, and A.I. Harris 1986, Ap. J. in press.

7. E. Serabyn, R. Gusten, J.E. Wink, C.M. Walmsby, and R. Zylka 1986, Astr. J., in press.

8. A.I. Harris, D.T. Jaffe, M. Silber, and R. Genzel 1985, Ap. J. <u>294</u>, L93.

9. K.Y. Lo 1986, Science Magazine, in press.

MAPPING OF C⁺ FAR-INFRARED EMISSION IN THE INNER GALAXY

J. B. Lugten, R. Genzel*, M. K. Crawford, and C. H. Townes
University of California, Berkeley, CA. 94720

ABSTRACT

We have mapped the [CII] 158 μm emission over a region of about 8' diameter towards Sgr A. The strongest C⁺ emission comes from a ring of gas centered on Sgr A*. The ring has an inner radius of about 1.7 pc and can be traced in C⁺ emission out to a least 8 pc from Sgr A*. The ring is inclined about 70° to the line of sight and tilted about -20° with respect to the galactic plane. The velocity field of the C⁺ is that expected for a tilted, inclined thin ring which is predominantly rotating about the center. There is also substantial "turbulent" velocity dispersion within the ring. The C⁺ rotation velocity decreases with distance from the center. If the velocity field is dominated by gravitational forces, the mass distribution in the center of the Galaxy is more concentrated than that of an isothermal stellar cluster.

OBSERVATIONS AND RESULTS

We have mapped the C⁺ 158 μm $^2P_{3/2} \rightarrow {}^2P_{1/2}$ fine structure line over a region of about 25 pc diameter toward the center of the Galaxy. In 1984, we observed the C⁺ line at 35 positions on a (Δl, Δb) grid; spectra were taken every arcmin on the grid with additional points near the center (open circles Figure 2b). The velocity resolution was 50 km s⁻¹ (FWHM) and the beam size was 55" (beam solid angle 9 × 10⁻⁸ sr). In 1985, we observed the C⁺ line at 10 positions

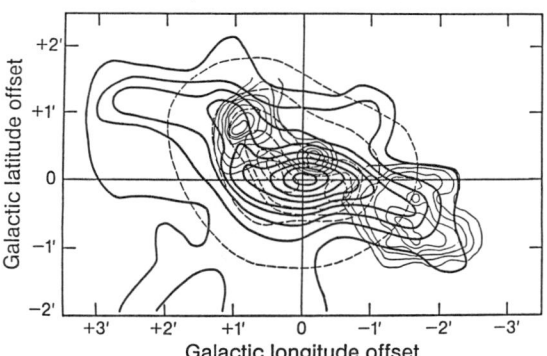

Figure 1: Maximum entropy deconvolved map of the integrated [CII] emission (heavy), superposed on a map of the integrated 63 μm [OI] $^3P_1 \rightarrow {}^3P_2$ emission[4] (dashed), and CO J = 1 → 0³, 80 < |v| < 110 km s⁻¹ (light).

*now at Max-Planck-Institut fur Astrophysik und Astrophysik, D-8046 Garching bei Munchen, Federal Republic of Germany

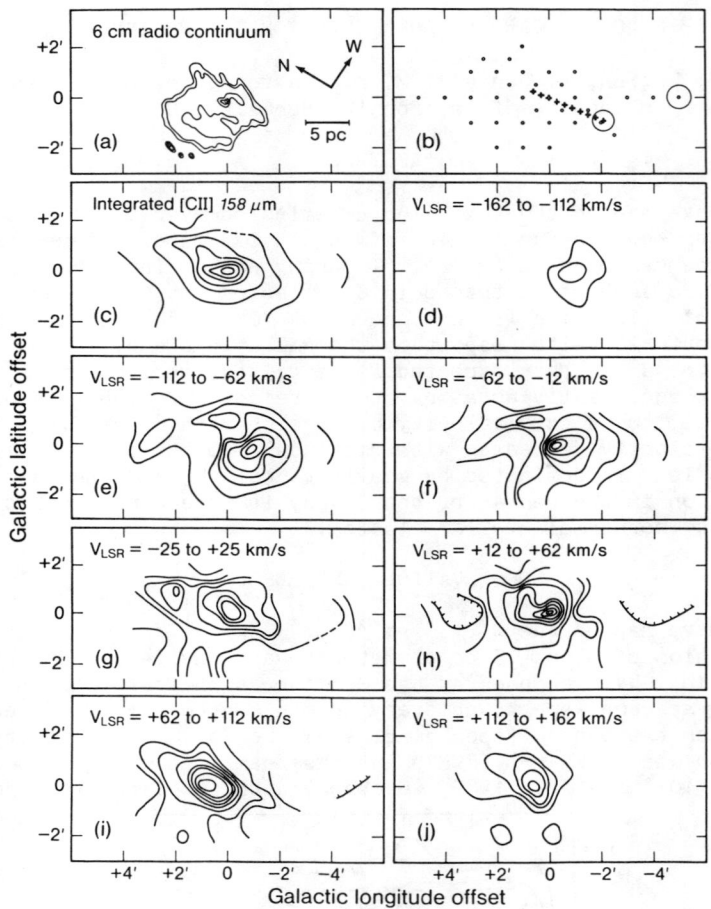

Figure 2: Maps of 158 μm $^2P_{3/2} \rightarrow {}^2P_{1/2}$ [CII] line emission and radio continuum emission in the galactic center. Coordinates are galactic longitude and latitude offsets with respect to the base position R.A. = $17^h 42^m 28.7^s$, Dec. = $-28° 59' 14"$ (1950). a) Schematic of 5 GHz radio continuum map[5] (beam size 5" × 8" (R.A. × Dec.)). b) Positions for [CII] observations. Dots mark the positions observed in 1984, crosses mark the positions observed in 1985. The FWHM beam sizes (marked by circles) were 55" in 1984 and 50" in 1985. Additional data were taken in 1985 toward 5 positions outside of the area plotted. c) Contour map of the integrated [CII] emission. The contour levels are linear with each contour representing 2×10^{-4} erg s^{-1} cm^{-2} sr^{-1}. d) through j) Linear contour maps of individual LSR velocity ranges. Each contour represents 4×10^{-5} erg s^{-1} cm^{-2} sr^{-1} per 50 km s^{-1} bandwidth. Maps at positive LSR velocities use eastern beam data. Maps at negative LSR velocities use the western beam data.

along a fully sampled strip (sampling 20", beam 50", beam solid angle 7 × 10⁻⁸ sr) at position angle -25° relative to the galactic plane (crosses in Figure 2b, along the major axis of the [CII] intensity distribution). The velocity resolution was 33 km s⁻¹ FWHM.

Based on the kinematic properties and spatial distribution of the C^+ emission, we conclude:

1) The strongest emission comes from a ring or thin disk with inner radius ~ 1.7 pc (assuming 10 kpc to the Galactic center) centered near Sgr A*. The ring emission is clearly detected to distances of about 8 pc from the center. The ring is tilted about -20° with respect to the Galactic plane and is inclined about 70° to the plane of the sky (see Figures 1 and 2).

2) The dominant motion of the ring is rotation (see Figures 3 and 4), with an additional "turbulent" velocity dispersion of 30 to 60 km s⁻¹ [1,2,3].

3) The C^+ spectra show rotational velocities which decrease substantially with distance from the center (Figure 3). Models for thin rings with turbulent velocity dispersions of 25 km s⁻¹ are shown in Figure 4. If the turbulent widths are broader near the inner edge of the molecular disk than at larger radius[1] then a somewhat slower falloff of rotational velocity with radius is implied. The masses shown in each panel of Figure 4 produce the indicated longitude-velocity behavior as long as gravitational forces dominate the gas motion.

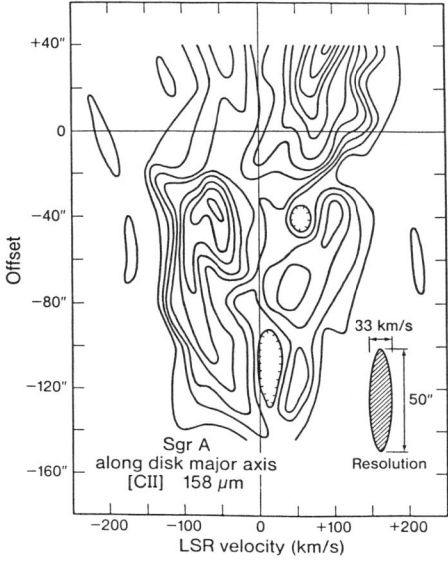

Figure 3: Longitude-Velocity (l-v) diagram of the 1985 [CII] data along the major axis of the [CII] intensity distribution (tilted 25° west of north relative to the galactic plane, see Fig. 1). Spectral and spatial resolutions were 33 km s⁻¹ and 50" (FWHM), and the data were spatially sampled every half beamwidth. For most spectra, the data from both beams were used.

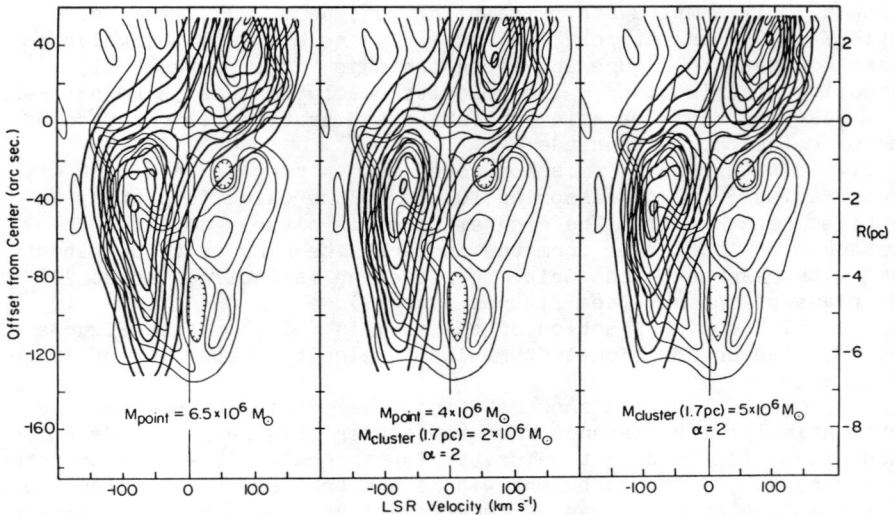

Figure 4: Example of rotating disk models (heavy lines) of the [CII] data along the major axis of the ring (thin lines, same as Fig. 3). In all three cases, a central cavity of $R_+ = 1.7$ pc without C^+ emission is assumed. At $R > 1.7$ pc the C^+ emissivity decreases with increasing radius. The turbulent width is taken to be $\Delta v_{FWHM} = 25$ km s^{-1} throughout the ring. Left: Central point mass of 6.5×10^6 M$_\odot$, no stellar cluster. Middle: Central point mass of 4×10^6 M$_\odot$, plus isothermal stellar cluster ($\rho_* \sim R^{-\alpha}$ with $\alpha = 2$): $M_{cluster}(R) = 1.2 \times 10^6$ R$_{pc}$. Right: Isothermal stellar cluster without central point mass. The stellar cluster has mass $M_{cluster}(R) = 3 \times 10^6$ R$_{pc}$.

We thank the staff of the Kuiper Airborne Observatory for their excellent support, and G. Stacey and W. Fitelson for their help in preparation of the spectrometer. This work was paid for by NASA grant NAG 2-208.

REFERENCES

1. R. Gusten, R. Genzel, M.C.H. Wright, D.T. Jaffe, J. Stutzki, and A.I. Harris, Ap. J., in press.

2. A.I. Harris, D.T. Jaffe, M. Silber, and R. Genzel, Ap. J. (Letters), **294**, L93 (1985).

3. E. Serabyn, R. Gusten, J.E. Wink, C.M. Walmsley, and R. Zylka, Astr. Ap., in press.

4. R. Genzel, D.M. Watson, M.K. Crawford, and C.H. Townes, Ap. J., **297**, 766 (1985)

5. R.D. Ekers, J.H. van Gorkom, U.J. Schwarz, and W.M. Goss, Astr. Ap. **122**, 143 (1983).

RADIO EMISSION FROM SGR A AND ITS EXTENDED HALO

Mark Morris
UCLA, Los Angeles, CA 90024

F. Yusef-Zadeh
Columbia Univ., New York, NY 10027 and UCLA, Los Angeles, CA 90024

ABSTRACT

Observations of the Sgr A complex with the VLA reveal that the diffuse structure surrounding Sgr A East and West consists of two extended components. One is a 20-pc-diameter halo which appears to be associated with the nonthermal Sgr A-E shell source, and which is elongated along the galactic plane. The second component is a set of narrow radio protrusions from the Sgr A-W complex which run perpendicular to the galactic plane. A few of these protrusions appear to merge smoothly with the arms of the inner "3-armed spiral" structure of Sgr A-W. These protrusions are discussed in the context of a model of a directed wind; magnetic fields may play some role in directing and collimating them.

Comparison of low and high-frequency maps reveals that the bulk of Sgr A-W is located in front of Sgr A-E. This supports the notion that Sgr A-E is unrelated to the activity in the galactic nucleus.

New, high-resolution images of the Sgr A-W complex show several new features of the inner few parsecs of the galaxy. Three of these are briefly discussed.

INTRODUCTION

In a recently submitted paper, we presented a VLA study of radio emission from Sgr A and its halo[1]. There, we argued that there are several features extended over $\geq$ 10 pc which might be physically linked to the activity evidenced within the inner few pc. These observations therefore have a bearing on the primary topic of this conference. We emphasize here a few of the points of that paper and present some preliminary results of ongoing VLA observations of the inner few pc.

LARGE-SCALE FEATURES IN AND AROUND SGR A

An example of an extended feature which may originate in the galactic nucleus is the low-frequency ridge of emission located at negative latitudes[2,3]. This ridge, observed between 110 and 160 MHz, runs ~ 30 pc perpendicular to the galactic plane and ends, or begins at Sgr A-W. It was suggested that this feature might represent a low-energy jet emanating from a compact central region.[2] If that hypothesis could be confirmed, it would be one of the strongest clues yet to the nature of the central object. High-resolution 327-MHz observations with the VLA can help to help settle this question.

Other features that originate in Sgr A-W and extend out to large distances can be seen at higher frequencies. Radiographs of Sgr A and its halo made at 6 and 20 cm are shown in figs 1 and 2,

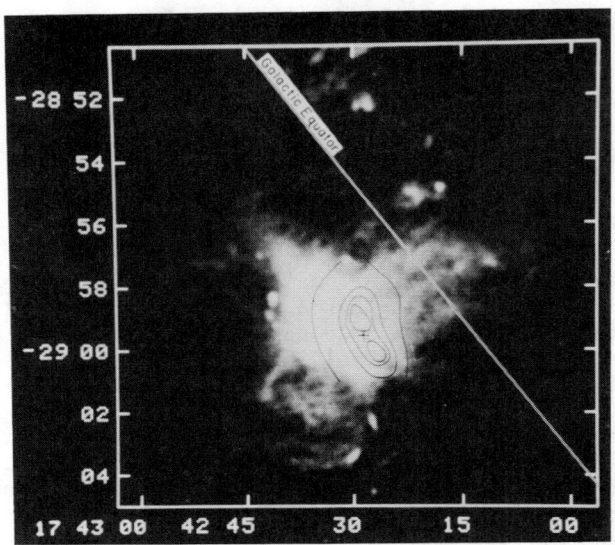

Figure 1: Contour map of 100μm emission from Sgr A-W[4], superimposed on a 20-cm VLA radiograph of the Sgr A complex.

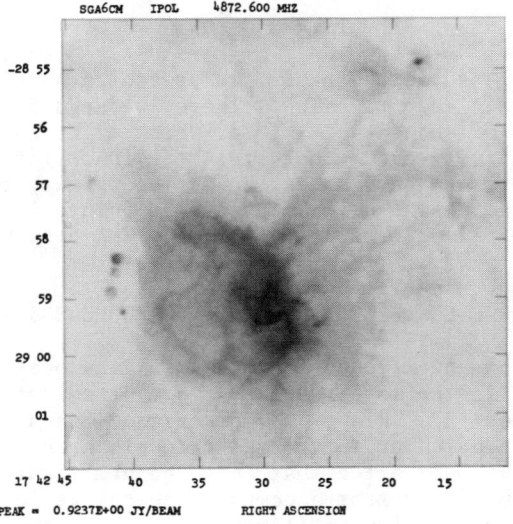

Figure 2: Radiograph of the Sgr A complex at a wavelength of 6 cm. The resolution is 3.4" x 2.9".

respectively. The most striking feature in both, outside the perimeter of the elliptical shell source Sgr A-E, is the set of streamers, or protrusions, that rise out of the vicinity of Sgr A-W and extend ~ 15 pc toward positive longitudes. All evidence (polarization, spectral index) currently indicates that these features are thermal in character. In fig. 1, some of these protrusions appear rather clearly to be extensions of linear structures within Sgr A-W, most notably the northern arm and the "bar". The whole set of protrusions converges toward the interior of the dusty molecular ring. This can be seen in figure 2; the superimposed contours of 100μm intensity[4] roughly define the ring. The protrusions appear to arise either from the inside edge of this ring, or from its interior.

One tempting model for the protruding streamers, inspired by the suggestion of a strong wind emanating from IRS16[5], is that a wind is being deflected out of the galactic plane by the ring. (Infall models encounter serious difficulties in accounting for the full extent of the protrusions, and the apparent continuity of the protrusions with features in the inner spiral.) However, there is no obvious negative-latitude counterpart to the protrusions, as might be expected in this "bipolar outflow" model. Curiously, the sense of the asymmetry in the protrusions is opposite to that of the low-frequency ridge.

If the streamers do result from a directed wind, what accounts for their filamentation? One possibility is that the wind is variable in time and direction, as some of the infrared results from this symposium have suggested. Each of the streamers would then represent a directional gust. The problem with this hypothesis is that the individual streamers diverge only slightly, and their surface brightness does not fall off with radius as much as might be expected in a diverging outflow. Something must act to keep the streamers collimated, and it is natural to invoke a magnetic field for that purpose. If the predominant magnetic field direction in the galactic center is poloidal, as we have suggested[1], and if the strength of the magnetic field is large enough for the field to be dynamically important (~ 10 milligauss[6,7]), then the field can be invoked to guide the outflow toward the galactic poles, and inhibit its divergence. With magnetic fields in the arena, one could identify the individual radio streamers with magnetic flux tubes which are placed in such a way that they are injected with relatively high amounts of the wind energy.

The persistence of the surface brightness of the streamers could be attributed to reheating of the wind material, perhaps by shocks which propagate radially (or along the field lines) in a manner analogous to those hypothesized for O-star winds.[8,9] A clear enhancement of X-rays occurs in the direction of the streamer-like protrusions from Sgr A.[1,10] As in O-star winds, the X-rays could be produced in shocks driven by radiation pressure.

Another large-scale radio structure which, because of its positional coincidence with the nucleus, has been considered as possibly relating to activity there, is Sgr A-E and its concentric, elliptical halo. In fact, the evidence now shows that the bulk of Sgr A-E lies behind Sgr A-W, and therefore that the two may be

Figure 3: Detail of Sgr A-W at 6 cm. Resolution: 0.55" x 0.28".

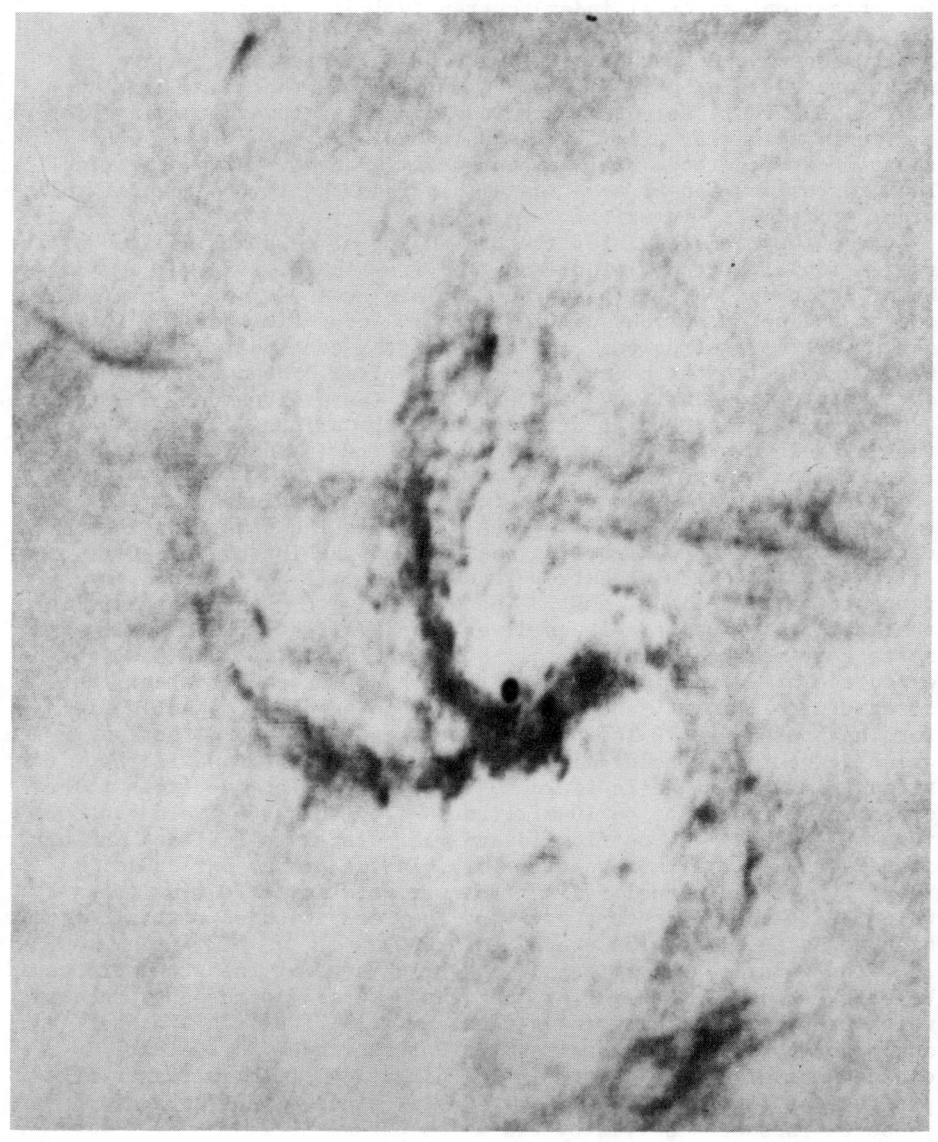

physically unrelated. Maps of 327-MHz emission show a pronounced depression at the position of Sgr A-W¹'¹¹, indicating that the background non-thermal emission from Sgr A-E has been absorbed in this thermal foreground source.

NEWLY-FOUND STRUCTURES IN THE INNER PARSEC

With R. Ekers, we have begun a detailed VLA study of Sgr A-W at high frequencies. Figure 3 shows the results of a 2-cm observation with the A configuration alone. The image is dominated by the compact source Sgr A*, which is clearly disconnected from the highly resolved emission from the "3-armed spiral". We point out the existence of two apparent "holes" in this picture. The first lies in the radio bar to the southwest of Sgr A*. This 0.1-pc hole has a bright western edge coinciding with IRS 2 and 13. We tentatively suggest that this hole is a cavity swept out by a wind or an explosive event. The bright edge is then explained as the highest-density portion of the hot compressed shell surrounding the presumably expanding bubble. The problem with this hypothesis is$_*$ that the obvious choice for the source of a wind -- IRS16 or SgrA* -- is not centered on the cavity. Perhaps the displacement reflects the inhomogeneity of the surrounding medium; that is, a relative absence of ambient material on the northern side of the source of the wind. Indeed, the high-density medium of the radio bar may be responsible for the absence of a negative-latitude counterpart to the large-scale streamers protruding from Sgr A. The other hole in the 2-cm picture (diameter ~ 0.2 pc) is centered at $\alpha=17^h42^m29.7^s$, $\delta=-28°59'22"$. It lies immediately to the East of the intersection of the northern and eastern arms of the radio spiral. Whether this feature is a cavity or an absorption patch is undecided.

Both of these holes can be perceived in figure 4, which is constructed from 6-cm data taken with the VLA in its "A" configuration. Because of the choice of transfer function, the cavity to the south of IRS16 is not as prominent as in the 2-cm map. It appears partially filled with emission.

Note also that with the high spatial resolution of figure 4, the southernmost portion of the northern arm is seen to have a sinuous character. If this can be shown to correspond to a transverse wave in the northern arm rather than to a fortuitous distribution of emitting and absorbing material, then it places an interesting constraint on models for the northern arm. For example, how could a tidally infalling stream give rise to such waves?

Finally, we point out the presence of a dark, linear feature in figure 4. Roughly parallel to the northern arm, and lying just to the west of it, is a dark band which stretches almost straight north from Sgr A*. This absorption feature (notice how it absorbs the faint east-west band of emission lying immediately above the center of the image) apparently lies in front of most of the emitting regions in this figure. But we suspect that it is not an unrelated foreground object, because it displays radio-bright rims, indicative of its being immersed in an intense UV radiation field. Perhaps there are more than three arms in the Sgr A-W spiral. Are we seeing the cool, optically thick side of one of them? The discussion of the temperature of this feature is deferred to a later publication.

REFERENCES

1. F. Yusef-Zadeh and M. Morris, Ap. J., submitted (1986).
2. F. Yusef-Zadeh, M. Morris, O.B. Slee and G.J. Nelson, Ap. J. (Letters), 300, L47 (1986).
3. N.E. Kassim, T.N. LaRosa, and W.C. Erickson, Nature, 332, 522 (1986).
4. E.E. Becklin, I. Gatley, and M. Werner, Ap. J., 258, 135, (1982).
5. T.R. Geballe, K. Krisciunas, T.J. Lee, I. Gatley, R. Wade, W.D. Duncan, R. Garden, and E.E. Becklin, Ap. J., 284, 118 (1984).
6. D.K. Aitken, P.F. Roche, J.A. Bailey, G.P. Briggs, J.H. Hough, and J.A. Thomas, M.N.R.A.S., 218, 363 (1986).
7. F. Yusef-Zadeh and M. Morris, this conference.
8. L.B. Lucy, and R.L. White, Ap. J., 241, 300 (1980).
9. L.B. Lucy, Ap. J., 255, 286.
10. M.G. Watson, R. Willingale, J.E. Grindlay and P. Hertz, Ap. J., 250, 142 (1981).
11. F. Yusef-Zadeh, M. Morris, O.B. Slee, and G.J. Nelson, Ap. J., 310 (1986).

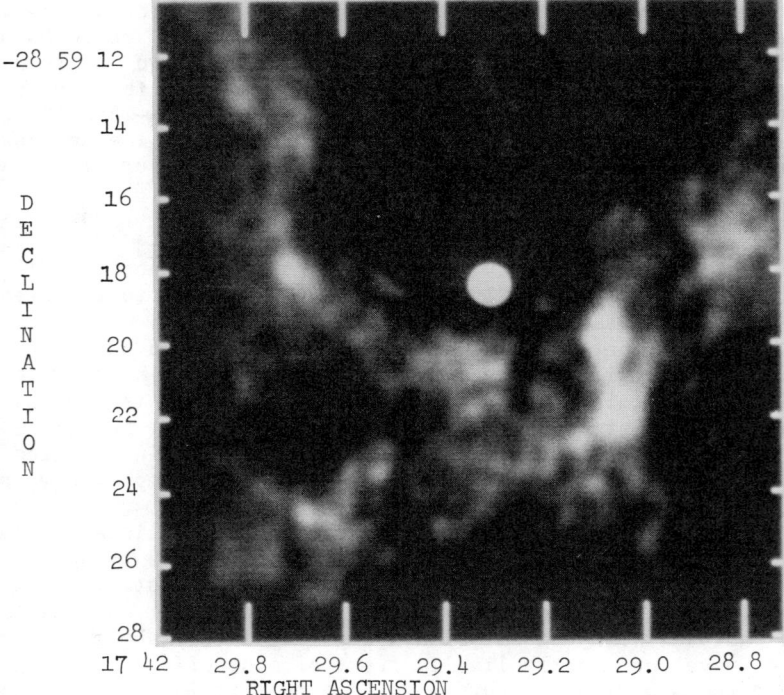

Figure 4: Radiograph of Sgr A-W at 2 cm. Resolution: 0.4" x 0.4".

86 GHz APERTURE SYNTHESIS OBSERVATIONS OF THE GALACTIC CENTER

M. C. H. Wright, R. Genzel, R. Güsten and D. T. Jaffe
(U. C. Berkeley)

ABSTRACT

We present 86 GHz aperture synthesis observations of the radio continuum with 4" x 8" angular resolution in the central 5 pc of the Galaxy. From a comparison with maps with comparable resolution at 1.4, 5, 15, and 24 GHz we have derived the distribution of thermal and non-thermal emission. The central radio source has a flux density 1.0 +/- 0.15 Jy between 1.4 and 86 GHz. The western arc is the inner ionised edge of the 2 pc radius molecular ring and can be traced around the western half of the ring. The central bar feature contains both thermal and non-thermal components. The apparent thermal spectral index between 1.4 and 5 GHz is due to a blend of non-thermal and optically thick thermal emission.

INTRODUCTION

The radio emission from the Galactic center has both thermal and non-thermal components. Sgr A East is a non-thermal ~2' diameter shell, which is probably a supernova remnant[1]. Sgr A West is a thermal radio source with spiral features which may be material falling into the galactic center[2]. Close to the center, ~1" from IRS16, is Sgra A*, a non-thermal point source, and possibly containing a black hole[3]. Surrounding all this activity is a ring of dust and neutral material[4]. At an assumed distance of 10 kpc, 1 pc subtends 20" and high resolution is required to seperate these phenomena. Using the Hat Creek mm-interferometer, we have mapped the central 5 pc in the HCN J=1-0 line at 88 GHz, the H42α line, and the radio continuum at 86 GHz. Here we compare the 86 GHz radio continuum observation with maps of comparable resolution at 1.4, 5, 15, and 24 GHz and derive the distribution of thermal and non-thermal emission.

OBSERVATIONS

Data were obtained between November 1985 and May 1986 with the Hat Creek mm-interferometer and 512 channel correlator. Observations were made with 15 configurations of 3 antennas in a single field, 2'.3 FWHM, centered on Sgr A*. The observations contain projected antenna spacings from 6 m. to 305 m. sampling structures between 1' and 2"; structures larger than ~45" are partially resolved. The instrumental gain and phase were calibrated by observations of the nearby extragalactic source NRAO530 at 40 min intervals. The flux density scale was established from observations of Venus. The synthesised maps were deconvolved using both the Clean and Maximum Entropy algorithms. The continuum map at 86 GHz is very similar to the maps at 5GHz, 15 GHz and 24 GHz (Figure 1). The features of these maps are discussed below.

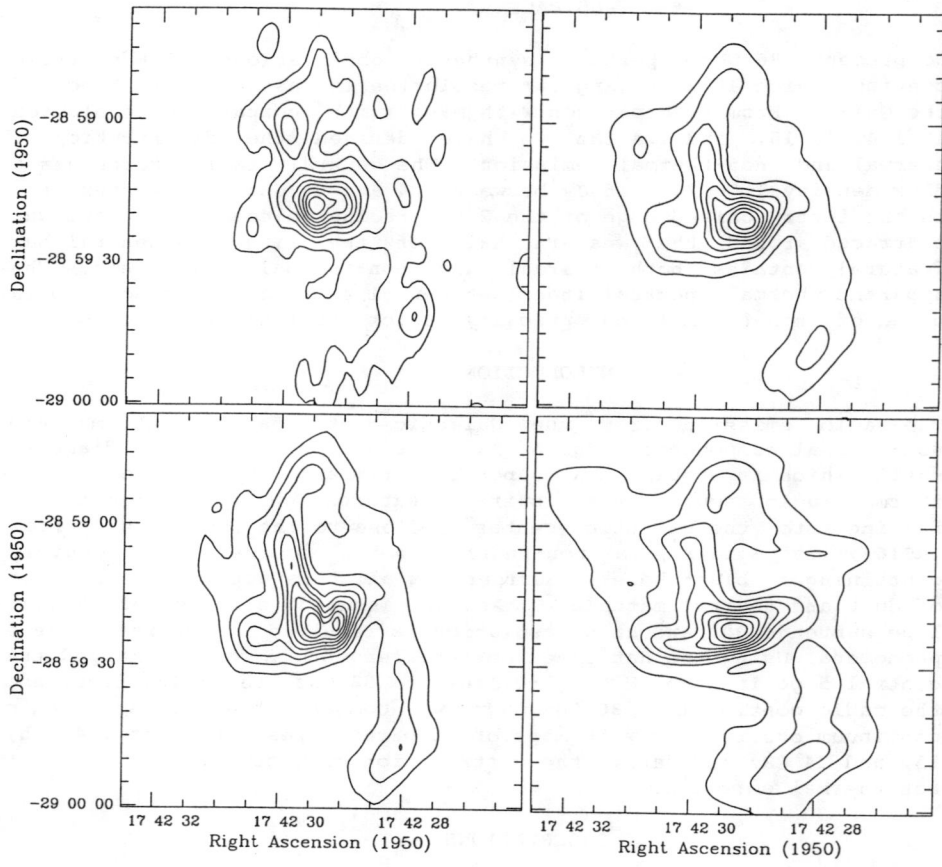

Figure 1. Continuum maps of Sgr A West convolved to a resolution 8".5 x 4".
Top left: 85.674 GHz, contours at interval of 88 mJy/beam.
Bottom left: 23.852 GHz, contour interval 89 mJy beam/beam. (ref. 5)
Top right: 14.697 GHz, contour interval 126 mJy/beam. (ref. 1)
Bottom right: 4.885 GHz, contour interval 184 mJy/beam. (ref. 2)

RESULTS

1 - compact nucleus

An unresolved source, whose position is consistent with the map center within 0".1, has been subtracted from each map in Figure 1. The flux density at 86 GHz was estimated from plots of the source visibility as a function of projected baseline. At projected spacings greater than 20000 λ the large scale structure is resolved out and the residual visibility is that of the compact source at the map center with flux density 1.05 +/- 0.15 Jy. The flux of the compact component at each frequency is given in table 1. Within the errors the flux from 1.5 to 86 GHz is constant.

2 - total flux

The total flux, corrected for the primary beamwidth was estimated by integrating the cleaned maps. The integrated fluxes in square boxes, after subtracting Sgr A^*, are listed in table 1. The 2 maps at 5 GHz are independent and give a measure of the errors.

Table 1 - Integrated fluxes

box	1.5 GHz	4.9 GHz	14.7 GHz	23.9 Ghz	85.7 GHz
1"	0.9	0.8 1.0	1.2	1.0	1.0
28"	7.4	10.8 9.8	9.4	7.1	5.5
56"	23.0	24.5 22.3	18.6	13.0	10.7
84"	40.0	33.7 30.3	22.8	14.1	12.6
refs.	1	1 2	1	5	

3 - continuum features

The western arc, which includes the southern arm and its extension to the north, the northern arm, and the east-west bar can be identified at each frequency. At 86 GHz the western arc can be traced as a ellipse of peaks extending to the north, and to the east beyond the southern tip of the distribution. At 5 GHz there is a background plateau of emission and the bar is more prominent, whilst at 1.4 GHz the background is much higher and a part of the southern arm is seen in absorption[1].

DISCUSSION

1 - spectral index

The integrated fluxes in table 1 are not consistent with a thermal spectrum. All the maps partially resolve structure larger than ~1' and the discussion which follows refers only to the structures visible on the maps in figure 1. We note, however, that single dish measurements: 14.5 Jy in a 60" beam at 15 GHz, 15 Jy in 42" at 22 GHz, 12 Jy in 38" at 43 GHz, and 10.5 Jy in 45" at 90 GHz, also contain non-thermal emission. The distribution of non-thermal emission at 5 and 15 GHz was estimated by subtracting maps at 86 GHz scaled by $\nu^{-0.1}$. The results at 5 and 15 GHz are similar. The non-thermal component consist of a large scale stucture enclosed by the neutral ring and has enhanced emission along the bar.

2 - western arc

The spectral index of the western arc is consistent with optically thin thermal emission, and this feature is probably the ionized inner edge of the molecular ring[4].

3 - northern arm

After allowing for the non-thermal background the northern arm is consistent with a thermal spectrum. Recombination lines are detected at 5, 15[7] and 86 GHz and velocities are consistent with a model of ionised gas falling into the Galactic center[2,6].

4 - bar

The apparent thermal index between 1.5 and 5 GHz can be understood as a blend of non-thermal and thermal emission with an optically thick component. The brightest clumps in the bar have brightness temperatures of 4000 K in a 1" beam at 5 GHz giving $\tau_{5GHz} \sim 0.4$ and $n_e \sim 10^5$ cm^{-3}. The presence of a non-thermal component also explains the failure to find recombination lines at 15[7] and 86 GHz. The dynamics of the bar are not well understood. Steep velocity gradients[6] and a discontinuity where the bar intersects the neutral ring preclude a gravitational explanation. Enhanced magnetic fields in the bar would not be altogether suprising in view of the flow of ionized plasma. Clumps of ionized gas in the bar have rotation measures to 10^5 rads m^{-2} and will depolarize background or entrained sychrotron emission; no polarization has been detected with a 1% upper limit at 15 GHz[8]. We might see regions of polarization at high angular resolution at high frequency, but this will depend on the geometry of the field.

The relationship of the non-thermal emission to Sgr A west is not clear. At points where the northern arm and bar intersect the neutral ring there are gaps in the HCN emission. (Figure 2). Non-thermal emission at 5 GHz appears to continue beyond these gaps. It appears that some of the non-thermal emission threads the ring and is connected to the inner arms. A flow of material away from the center , directed along magnetic field lines would provide a mechanism for removing angular momentum from the infalling material along the northern arm. Magnetic fields could also provide some support for the sharp inner edge of the neutral ring seen in HCN emission.

REFERENCES

1. R. D. Ekers, J. H. van Gorkom, U. J. Schwartz, and W. M. Goss, Astron. and Astrophys., 122, 143. (1983)
2. K. Y. Lo and M. J. Claussen, Nature, 306, 647. (1983)
3. M. J. Rees. This volume.
4. R. Guesten. This volume.
5. P. T. P. Ho and J. Jackson. in preparation.
6. E. Serabyn and J. H. Lacy, Ap. J. 293, 445. (1985)
7. J. H. van Gorkom, U. J. Schwartz, and J. D. Bregman, I. A. U. Symposium No. 106, p 371, The Milky Way Galaxy, eds. H. van Woerden et al. (Reidel, Dordrecht, 1985)
8. J. H. van Gorkom, private communication.

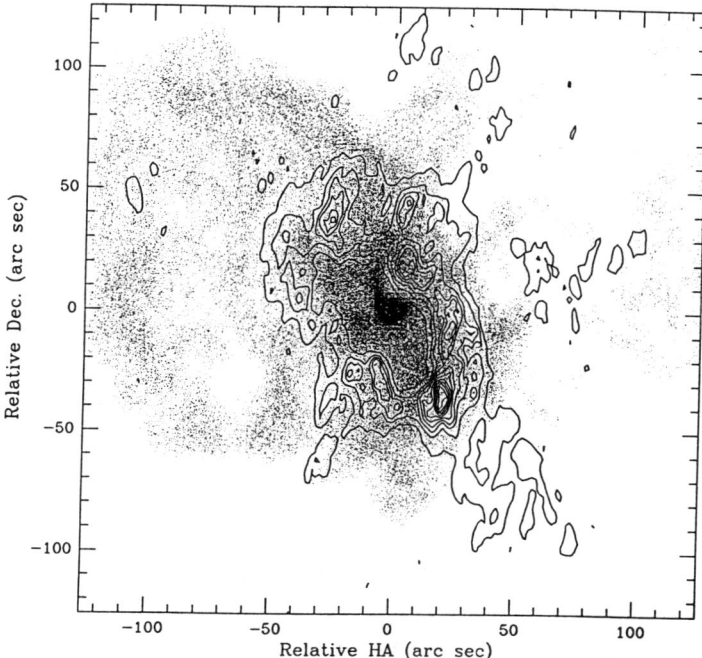

Figure 2. Contours of the HCN emission (contour interval 0.15 K averaged over 300 km s^{-1}.) , supperposed on the 5 GHz continuum (ref. 1)

SMALL SCALE STRUCTURE OF THE GALACTIC CENTER IN THE FAR-IR

D. F. Lester, M. Joy, P. M. Harvey and H. B. Ellis Jr.
Department of Astronomy and McDonald Observatory, University of Texas

ABSTRACT

Far infrared slit scans across the Galactic center are presented. These scans give spatial information on the luminosity distribution of this region on the smallest scale yet attained. These scans, made along the galactic plane through IRS16, show two luminosity peaks 39" (2 pc) apart. The NE peak is considerably hotter than that to the SW. When image restoration and deconvolution techniques are applied to the data, it is found that the luminosity peaks are surrounded by a region of cold dust emission. The cool dust emission appears to coincide spatially with the molecular ring, while the luminosity peaks are completely contained within the ring.

INTRODUCTION

The Galactic center is a strong source of far infrared continuum emission. This radiation is predominantly thermal emission from dust grains that are intermixed with the gas, heated by the intense stellar radiation field. The far infrared continuum thus can act as a tracer of the dust column density and the ambient energy density.

This paper presents first results of a high spatial resolution study of the Galactic center in the far infrared. This work was made possible using special scanning and tracking techniques on the Kuiper Airborne Observatory that are described in detail in recent papers.[1,2,3] Image restoration methods were applied to the data to make full use of the information available in the slit profiles. Slit scans parallel and perpendicular to the Galactic plane yield information on spatial scales as small as 10" when deconvolved with the appropriate point source profiles. This is a factor of two to three smaller than that previously available at these wavelengths.[4,5]

OBSERVATIONS

The observations were made at an altitude of 41000 feet on 23 April during the 1986 southern hemisphere deployment of the KAO to Christchurch, New Zealand. The spatial scans were made with slits of width $\sim \lambda/D$ (12" at 50μm and 24" at 100μm) and length 35". The point source profile (PSP) was determined from scans of Uranus (1.9" disk) taken on the same flight. A representative subset of the raw data (7 scans each, shifted vertically for comparison) is shown in Figure 1a and 1b.

The two far infrared lobes that were first detected in earlier studies[4,5] are well resolved in every scan. It is also apparent that the shape of the profile is quite different at the two wavelengths. The scans were subsequently low pass filtered to remove frequencies above D/λ and shifted slightly in order to correct for field rotation induced shifts. The resulting coadded profiles, with $\pm 1\sigma$ error bars on each bin, are shown in Figures 2a and 2b. Superimposed on these, and normalized to the same peak, is the PSP at the appropriate wavelength. The PSP is centered on the bin position corresponding to IRS 1/16. The S/N on the PSP is clearly the limiting factor in the deconvolution.

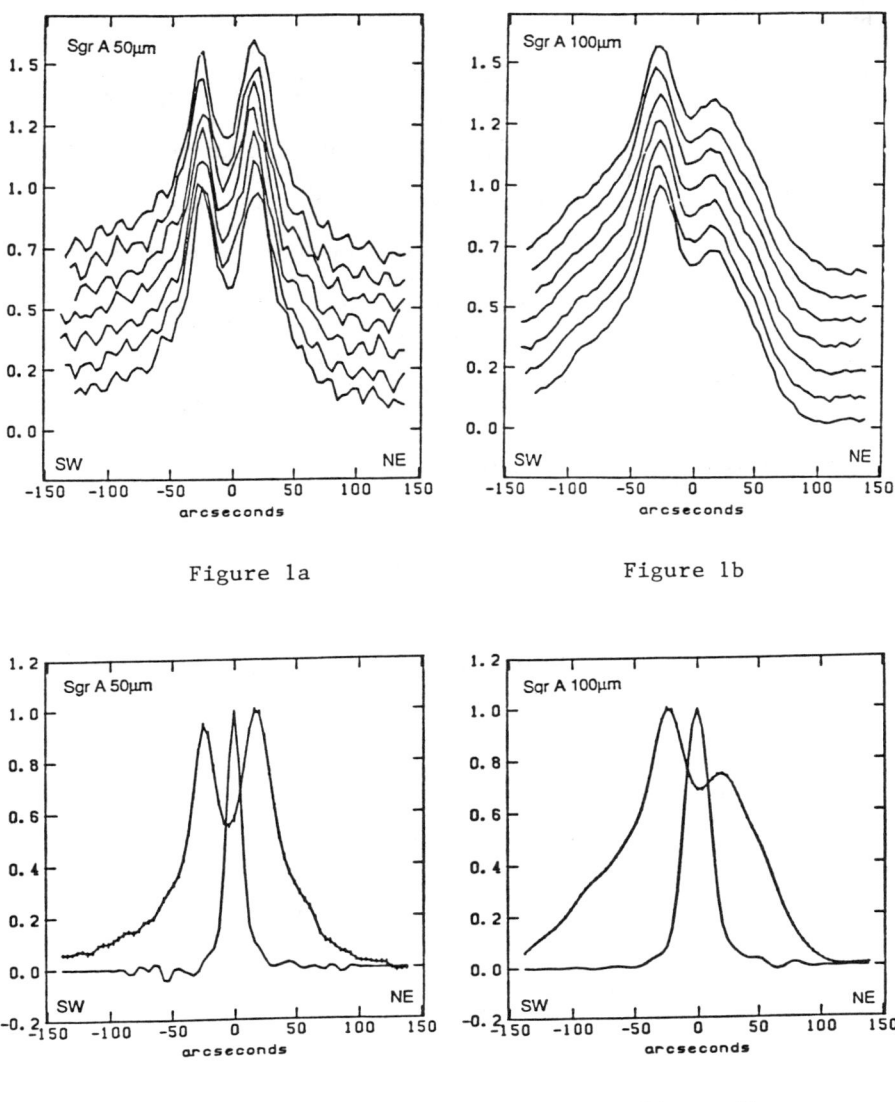

Figure 1a

Figure 1b

Figure 2a

Figure 2b

SMALL SCALE FAR INFRARED PROPERTIES OF SGR A

Comparison with the PSP indicates that each of the far infrared lobes is well resolved. Deconvolution of the data gives a FWHM of approximately 25" for each lobe at 50µm and there is evidence for a shoulder on each side of the profiles at a distance of 50-70" from the center. The positions of highest surface brightness in the lobes are separated by 39" at this wavelength. This is considerably smaller than the diameter of the molecular disk[6], which evidently surrounds the far infrared lobes. The two lobes appear to have rather different color temperature; the NE lobe is substantially hotter than that to the SW. A more detailed comparison of the profiles at the two wavelengths is made possible by the Fourier beammatching.[2] This results in a 50µm profile that is reconvolved to that appropriate to the 100µm beam profile. The 100µm profile and the beammatched 50µm profile are shown superimposed in Figure 3.

Figure 3 shows, in addition to the temperature differences noted above for the lobes, that the shoulders on the sides of the profile are very cool. While dust temperatures of ~60K predominate in the core, these shoulders indicate that the surrounding dust is only at ~40K (assuming $1/\lambda$ dust emissivity law). When compared with a consistent PSP, the 100µm emission remains broader than the 50µm emission, although the separation of the peaks is similar at the two wavelengths. It is tempting to associate this cool material at 50-70" radius with the molecular ring that is seen in HCN.[6] In view of the slit size and the scan path, the greater extent of the infrared emission in the SW direction resembles that seen in the molecular maps.

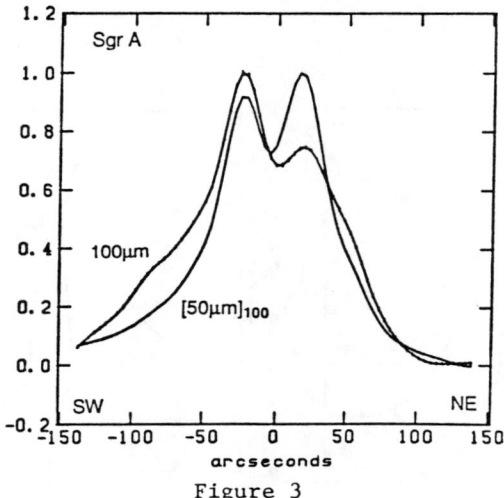

Figure 3

In Figure 4 we show the result of maximum entropy deconvolution of the data in Figure 2. The results for 50 and 100μm are superimposed in the Figure, and the two deconvolved scans are compared with the HCN and radio continuum map (reproduced from Gusten et al., reduced to the same scale and rotated to match the position angle of the far infrared scans). The cool wings on the side of the profile are evident in this presentation, as in Figure 3. The deconvolution suggests that the far infrared emission is strongly concentrated in the peaks, which sit on top of a broader, lower surface brightness pedestal. The spatial extent of this relatively cool emission pedestal is similar to that of the molecular ring, and we identify this component with that gas. The far infrared peaks are completely contained within the ring structure. We may speculate that the emission peaks are related to the heated inner edge of the ring, though the spatial scales are somewhat different.

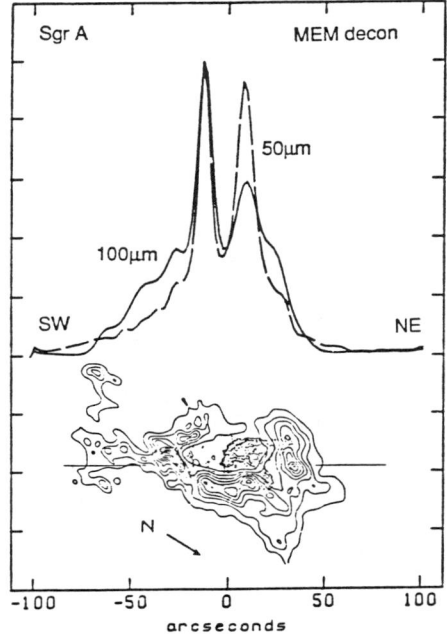

Figure 4

MEM decon

HCN, HII

This work was supported by NASA Grant NAG2-67 to the University of Texas.

REFERENCES

[1] Lester, D.F., Harvey, P.M., and Joy, M. 1986 Ap.J. 302, 208.
[2] Lester, D.F., Harvey, P.M., Joy, M. and Ellis, H.B. 1986 Ap.J in press.
[3] Wilking B. A. et al. 1984, Ap. J. 279, 291.
[4] Becklin, E.E., Gatley, I., and Werner, M.W. 1982 Ap.J. 258, 135.
[5] Harvey, P.M., Campbell, M.F., and Hoffmann, W.F. 1976 Ap.J. 205, L69. (see also ApJ Letters 241,L183)
[6] Gusten, R. et al . 1986 preprint.

OBSERVATIONS OF GALACTIC CENTER GAS DYNAMICS WITH A CRYOGENIC ECHELLE SPECTROMETER

J. H. Lacy, D. F. Lester
Department of Astronomy and McDonald Observatory, University of Texas

J. F. Arens, M C. Peck
Space Sciences Laboratory, University of California, Berkeley

and

S. Gaalema, Hughes Aircraft, Carlsbad California

ABSTRACT

Spectra have been obtained of the [Ne II] (12.8 μm) line from the central 12" x 14" of the Galaxy with a new cryogenic echelle spectrometer. The central "bar" region of Sgr A West was completely mapped with ~ 2" and 30 km s^{-1} resolution by measuring spectra in 7 slit positions with a 2" x 12" slit with pixels each 1.2" along the slit. The improved sensitivity and simultaneous spectral and spatial sampling resulting from the use of a 10 x 64 element detector array provide new insight into the flows of gas in the Galactic Center.

INSTRUMENT

The data presented here are the first obtained with a new liquid helium cooled echelle spectrometer which uses a 10 x 64 element Si:Ga detector array. The instrument contains a 10 x 22.5 cm echelle in a .75m f.l., f/7.5 spectrometer. The entrance slit, which was set at 2" width, is projected onto the short dimension of the array with 10 pixels covering 12" when used on the McDonald Observatory 2.7m telescope. The spectrum is dispersed along the 64 pixel length of the array with a dispersion of 18 km s^{-1}/pixel and a resolution of ~ 30 km s^{-1}. Diffraction limited spatial and spectral (0.03 cm^{-1}) resolution should be obtainable. The measured NEFD was 40 Jy/$\sqrt{\text{Hz}}$ or NELF = 6 x 10^{-16} W/m^2/$\sqrt{\text{Hz}}$ for each 1.2" x 2" pixel. For these observations, the data acquisition system of the U.C.B. Infrared Camera (Arens et al[1]) was used. It provides programmable array clocking and low-noise A/D conversion, resulting in photon-noise-limited performance of the system.

OBSERVATIONS

The data were taken with the McDonald Observatory 2.7m telescope on the night of 20 July 1986. The [Ne II](12.8μm) line was observed with -675 to +475 km s^{-1} Doppler coverage and 30 km s^{-1} resolution. Seven slit positions were measured in the central "bar" region of Sgr A West and one position was measured along the "northern arm". The 2" x 12" slit was oriented north-south. In the bar region, spectra were measured each 2" or 0.^{s}15 between 17^h 42^m 29.^{s}05 and 29.^{s}95, with the slit center at -28° 59' 22". The northern arm slit center was 17^h 42^m 29.^{s}80, -28° 59' 14".

The combination of seeing (~ 1"), diffraction, pixel size, and optical aberrations resulted in a spatial resolution ~ 4". To recover the degraded spatial information, the data were processed with a maximum entropy method deconvolution routine, which resulted in ~ 2" resolution. The deconvolution procedure had little effect on spectral structure, as the line components were generally well resolved. No new features were created by the deconvolution.

The MEM deconvolved data are shown in Figures 1 and 2. Each frame corresponds to one slit position with the spectrum displayed vertically and declination horizontally. Continuum emission has been subtracted from the spectra and only the central ± 400 km s^{-1} are shown. A spectrum from a location 4" south of Sgr A* is shown in Figure 3.

DISCUSSION

Strong line emission is seen over the range ± 400 km s^{-1}, with evidence for emission at -600 km s^{-1} south of Sgr A* (Figure 3). The observed extent of the line emission (at least 800 km s^{-1} FWZI) is larger than had previously been recognized and approaches that of the He I 2.06 µm line (Hall et al.[2] and Geballe et al.[3]), particularly if the -600 km s^{-1} emission is real. It would be very difficult to account for such large velocitries by infall unless a compact massive object dominates the potential. However, the highest velocities appear not to be localized at the presumed center (IRS 16 or Sgr A*), but a few arcseconds south. Confirmation of the ~ 600 km s^{-1} emission is required.

Several apparently connected features can be seen in the data:

1) The "northern arm" runs north from IRS 1 (29.^{s}8, 19") past IRS 10 (29.^{s}8, 14") with a velocity gradient of $\Delta v/\Delta \delta$ = 11 km s^{-1} arcsec^{-1} (Figure 2).

2) A ridge runs SW from IRS 1, then W to IRS 2 (29.05, 23") with the velocity changing from ~ 0 km s^{-1} to ~ -300 km s^{-1}, with $\Delta v/\Delta \alpha$ = 30 km s^{-1} arcsec^{-1}. The data strongly suggest that this feature connects physically with the northern arm, with an abrupt change in direction and velocity gradient at IRS 1. This interpretation was already suggested by previous [Ne II] data and is discussed by Serabyn, Lacy, and Townes (in preparation).

3) A spatially extended feature is seen centered near 29.^{s}20, 22" with a velocity near +20 km s^{-1}. Its small velocity gradient suggests that it is located ~ 1 pc from the center and is seen in projection toward the center.

4) A very broad line (+ 100 to + 400 km s^{-1}) is seen toward IRS 9 (29.^{s}65, 26"). This feature may be due to: several velocity components along the line of sight; a collision between two gas flows; or an expanding shell centered near a minimum in the VLA image at 29.^{s}75, 23". In the last case, the broken rings toward the tops of the plots near 29.^{s}65 would represent one connected feature.

5) A bridge apparently connects IRS 9 with IRS 5 (29.80, 09") in Figure 2. This would seem to imply a splitting of the northern arm along the line of sight.

The present data strongly support the view that much of the ionized gas in the Galactic Center moves along organized flows (e.g., Lo and Clausen[4] and Serabyn and Lacy[5]). Serabyn and Lacy show that several of the flows are well explained by unperturbed orbits in a gravitational potential. It now appears that in a few localized regions substantial deflections of the flows occur. In addition, some organized structures may require other explanations, such as explosions.

ACKNOWLEDGEMENTS

This work was supported by NSF grant AST85-02401 and the U. T. University Research Institute. We thank M. Huang for assistance with the observations, and E. Serabyn and C. H. Townes for helpful discussion and encouragement.

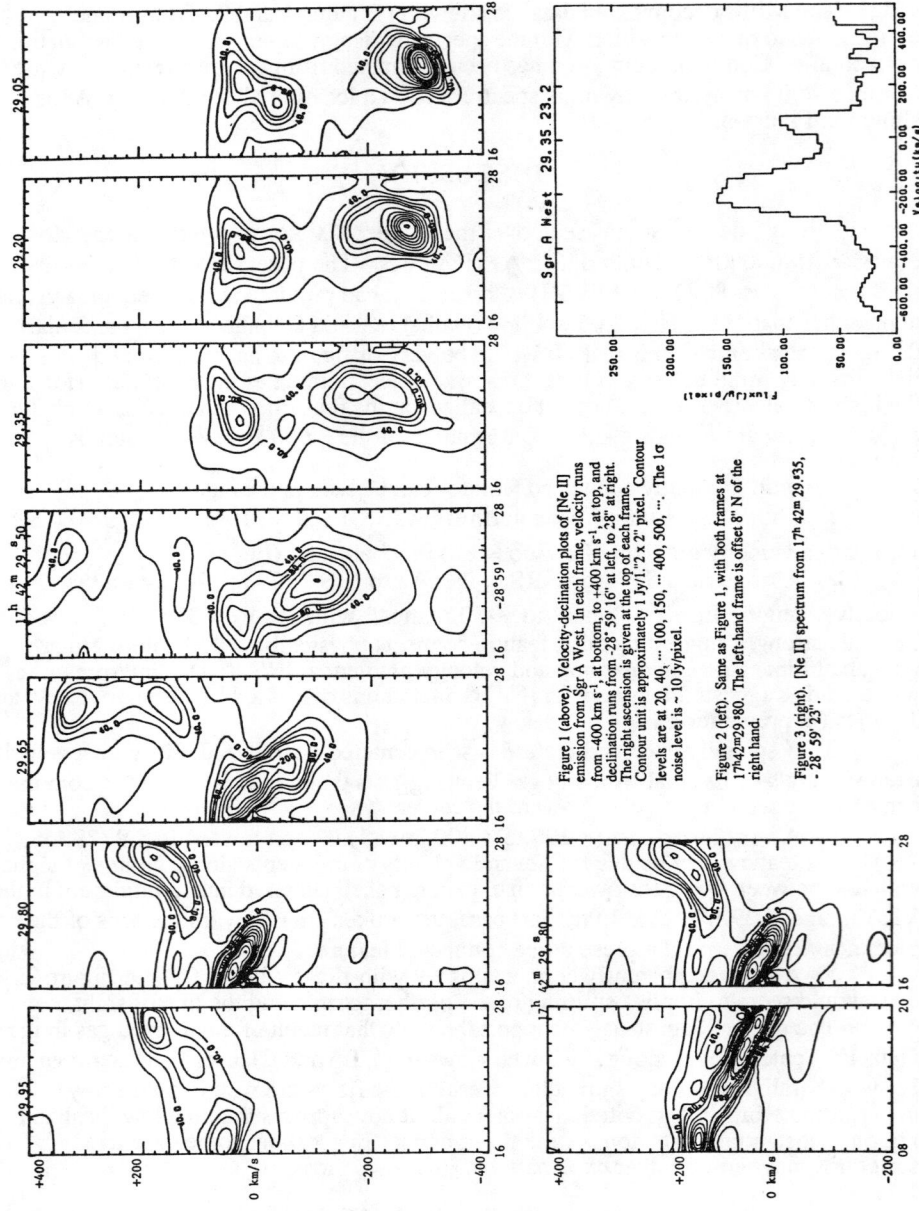

Figure 1 (above). Velocity-declination plots of [Ne II] emission from Sgr A West. In each frame, velocity runs from -400 km s^{-1}, at bottom, to +400 km s^{-1}, at top, and declination runs from -28° 59' 16" at left, to 28" at right. The right ascension is given at the top of each frame. Contour unit is approximately 1 Jy/1."2 x 2" pixel. Contour levels are at 20, 40, ··· 100, 150, ··· 400, 500, ···. The 1σ noise level is ~ 10 Jy/pixel.

Figure 2 (left). Same as Figure 1, with both frames at 17^{h}42^{m}29.80. The left-hand frame is offset 8" N of the right hand.

Figure 3 (right). [Ne II] spectrum from 17^h 42^m 29.835, -28° 59' 23".

REFERENCES

1. J. F. Arens, J. G. Jernigan, M. C. Peck, C. Dobson, E. Kilk, J. H. Lacy, and S. Gaalema, in preparation.
2. D. N. B. Hall, S. G. Kleinman, and N. Z. Scoville, *Ap. J. (Letters)*, **260**, 653 (1982).
3. T. R. Geballe, *et al. Ap. J.*, **284**, 118 (1984).
4. K. Y. Lo, and M. J. Claussen, *Nature*, **306**, 647 (1985).
5. E. Serabyn, and J. H. Lacy, *Ap. J.*, **293**, 445 (1985).

8.3 and 12.4 MICRON IMAGING OF THE GALACTIC CENTER WITH THE GODDARD INFRARED ARRAY CAMERA

D. Y. Gezari[1], R. Tresch-Fienberg[2], G. G. Fazio[2], W. F. Hoffmann[3], I. Gatley[4], G. Lamb[1], P. Shu[1], and C. McCreight[5]

A 30 x 30 arcsec field at the Galactic Center (1.5 x 1.5 parsec) has been mapped at 8.3μm and 12.4μm with high spatial resolution and accurate relative astrometry using the 16 x 16 Si:Bi AMCID Goddard infrared array camera. A 3σ upper limit of 0.5 Jy/arcsec^2 was set for the strong near infrared source IRS 16 at the nominal Galactic Center (Sgr A*). Color temperature and dust opacity distributions derived from the spatially registered 8.3μm and 12.4μm images show that the compact infrared sources are essentially density features. IRS 1, 5 and 10 are only slightly warmer (T_c = 290K) than the surrounding material; IRS 2, 4, 6 and 9 are indistinguishable in color temperature from the extended cloud complex (T_c = 260K). These observational results and interpretation are discussed in greater detail by Gezari et al. (1985).

Evidence is accumulating which suggests the presence of a massive luminous object at the Galactic Center. 30-100μm observations suggest that a gas- and dust-depleted region several parsecs in diameter centered on Sgr A* is swept clean by a central luminosity source which powers the far-infrared emission (Becklin, Gatley, and Werner 1982). The velocity distribution observed by Lacy et al. (1980) and Serabyn and Lacy (1985) in the 12.8 μm Ne II line, the CO work by Harris et al. (1985) and the far infrared line observations of Genzel et al. (1985) all are consistent with the presence of a massive central object. Reviews by Gatley and Becklin (1981) and Brown and Liszt (1984) discuss the possible role of a central object in explaining the energetics of the innermost few parsecs and the surrounding region of diameter ∼ 100 pc.

The Goddard array camera contains an Aerojet ElectroSystems Co. 16 x 16 pixel Si:Bi AMCID (bismuth doped silicon accumulation mode charge injection device) monolithic array detector with 254 active pixels. The characteristics and performance of Si:Bi AMCID array detectors have been discussed in general by Parry (1980, 1983), McCreight and Goebel (1981), and McKelvey et al. (1985). A detailed description of the 16 x 16 array camera system used here is also given by Lamb et al. (1984).

[1] NASA/Goddard Space Flight Center
[2] Harvard-Smithsonian Center for Astrophysics
[3] Steward Observatory, University of Arizona
[4] United Kingdom Infrared Telescope
[5] NASA/Ames Research Center

The Galactic Center observations were made on the nights of 1983 August 16 and 17 at the 3-meter NASA Infrared Telescope Facility (IRTF) at Mauna Kea, Hawaii. Beam switching was 30 arcsec to the south at 10 Hz. The optical plate scale at the IRTF is 0.78 arcsec per detector pixel. The observed beam profile (FWHM) at both wavelengths was 1.3 arcsec in declination and 2.3 arcsec in right ascension, including tracking and infrared focus errors, determined from images of β Peg. The 1σ noise equivalent flux density (NEFD) of the camera system during operation with typically 5% bandwidth filters was NEFD ∿ 0.1 Jy per pixel in a 60 second integration, the limit for extended sources. Since a point source image is spread over several detector pixels, the minimum noise equivalent point source flux density is about 2 Jy in 60 seconds.

Table 1: 16 X 16 Si:Bi AMCID Array Camera System Characteristics

Well Capacity	$= 2 \times 10^5$ electrons
Read Noise[a]	< 600 electrons
Dark Current	$= 200$ electrons sec^{-1}
Quantum efficiency	$\lesssim 5\%$
Responsivity	$= 0.5$ Amp W^{-1} (high background)
Noise Equivalent Power[b]	$= 4 \times 10^{-17} W Hz^{-1/2}$ (zero background)
Observational NEFD	$= 0.1$ Jy per pixel in 60 seconds
Optical Frequency Response	$= 200$ Hz (@ 10^9 photons $sec^{-1} pixel^{-1}$)
	$= 25$ Hz (@ 10^7 " " ")
Maximum Frame Rate	$= 800$ Hz

[a] System noise limited. 150 electrons detector read noise measured by Ames (2 x 64 Si:Bi array), <100 electrons for this family of detectors estimated by Aerojet.

[b] Considering noise derived from dark current only (zero background).

Most of the strong compact 10μm sources in the Galactic Center region are not prominent features in the derived color temperature distribution (Figure 2) calculated from the ratio of 8.3μm/12.4μm emission (Figure 1). We find the curved ridge of 10μm emission corresponding to the northern arm and ionized bar to be a region of almost constant temperature, and cooler than the surrounding diffuse material. The color temperature results show that temperatures away from the arm and bar are generally higher than temperatures within the complex of 10μm sources, although this result derives from data in regions with the lowest signal/noise. IRS 1, 5 and 10 are only slightly warmer ($T_c = 290K$) than the surrounding material; IRS 2, 4, 6 and 9 are indistinguishable in color temperature from the extended

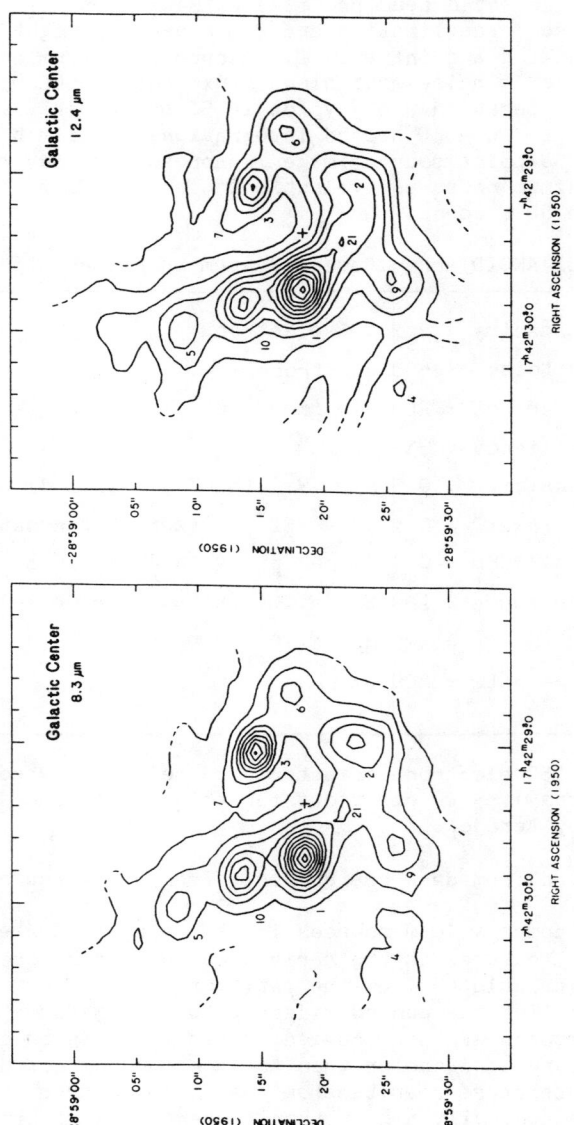

Figure 1 (a) and (b): Calibrated surface brightness isophotes for the Galactic Center region at 8.3μm and 12.4μm. The lowest contour plotted is 3σ above the noise level. Compact sources are labeled with their IRS numbers. The position for IRS 16(Center) and Sgr A* is shown by the cross. (a) The 8.3μm brightness distribution; the contour interval is 0.61 Jy/arcsec^2, a 6σ increment. The peak 8.3μm brightness is 7.0 Jy/arcsec^2 at IRS 1. (b) The 12.4μm brightness distribution; the contour interval is 1.36 Jy/arcsec^2, a 6σ increment. Peak brightness is 16.9 Jy/arcsec^2 at IRS 1.

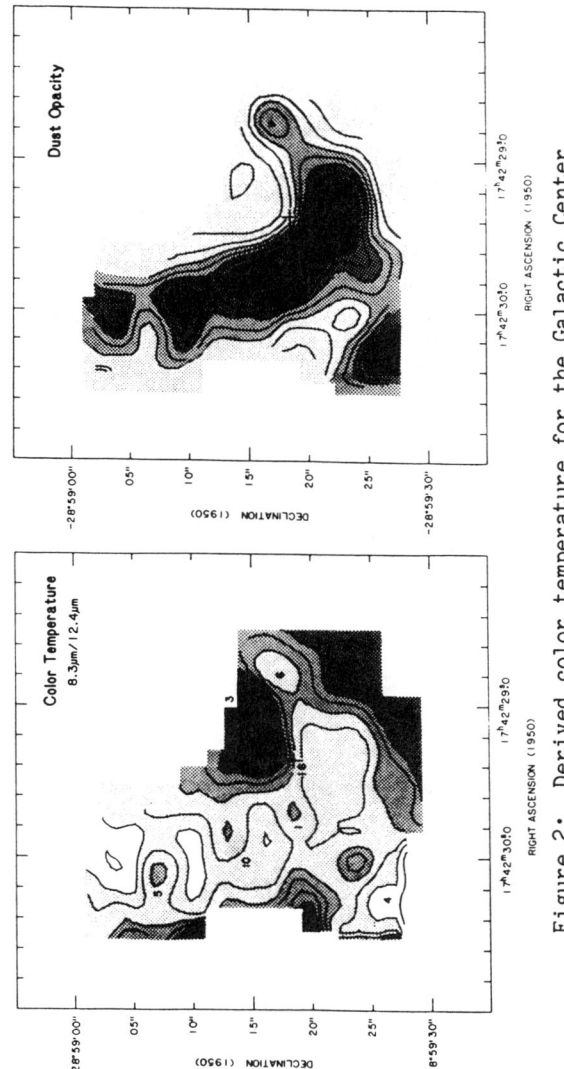

Figure 2: Derived color temperature for the Galactic Center source complex, calculated from the 8.3μm/12.4μm image ratio assuming blackbody spectra. Darker shades are higher temperature. The position of Sgr A* and IRS 16(Center) is shown by the cross. The lowest temperature contour is T_c = 220K; contours are plotted at 20K intervals (1σ = 5K in regions of high infrared surface brightness). The peak color temperature T = 400K occurs at IRS 3. IRS 1, 5 and 10 are each 290K, and IRS 6 is 260K. Higher color temperatures based on lower signal to noise data in the weakest regions may not be physically significant (see Figure 3).

Figure 3: Derived opacity of warm dust grains giving rise to the observed Galactic Center 8.3μm emission. Darker shades are high opacity. The position of IRS 16(Center) is shown by the cross. The lowest contour plotted is τ = 5.0 x 10^{-4}, and the contour interval is τ = 5.0 x 10^{-4}. The peak opacity τ = 5.6 x 10^{-3} occurs at IRS 1.

cloud complex (T_c = 260K). The significant implication of these results is that the 10µm IRS sources are not dominant factors in the large scale temperature structure of the region.

Previous workers have identified the Galactic Center 10µm sources as local temperature peaks (Rieke and Lebofsky 1982; Aitken, Allen and Roche 1982; Rieke, Telesco, and Harper 1978; and Becklin et al. 1978), and the presence of a compact infrared source at a local peak in color temperature has been frequently cited as evidence for internal heating by an embedded luminosity source, possibly a recently formed star. It has been suggested that the near infrared emission from these sources arises from a central star, and the 10µm emission from the surrounding dust cloud (Lebofsky et al. 1982). In the high resolution, spatially calibrated array camera images presented here, only the IRS sources in the northern arm appear as local temperature maxima. The others (not including the stars IRS 3 and 7) do not exhibit the temperature structure expected of objects containing embedded luminosity sources. The maps of derived color temperature and dust emission optical depth imply that the 10µm complex is bright not because it is self luminous but because it represents a significant enhancement in dust density.

The observed 8 - 13µm emission from the Galactic Center source complex is optically thin everywhere with a typical value of $\tau = (2 \pm 1) \times 10^{-3}$. Thus, infrared radiation is detected with equal efficiency throughout the central few parsecs of the Galaxy. The northern arm and ionized bar are clearly seen as dust density enhancements. The opacity is greatest near all of the 10µm objects, and falls off steeply away from them. Even relatively weak sources like IRS 4, 5, and 6 are strong density peaks. The lowest densities in Figure 5 occur in the region of apparently enhanced color temperature off the ridge between IRS 1 and IRS 3. IRS 3 itself is barely discernible as a density feature, a manifestation of its relatively compact circumstellar dust shell.

Using the standard dust/gas mass ratio (0.01), we derive an ionized gas mass of $\sim$ 50 $M_\odot$ in the central parsec, consistent with the results of VLA observations of the region by Lo and Claussen (1983). Since Harris et al. (1985) and Genzel et al. (1985) deduce a mass of $\sim$ 10^4 $M_\odot$ in the $r \sim 2$ pc ring, our results support the contention of Becklin, Gatley, and Werner (1982) that the volume immediately surrounding the Galactic Center is significantly depleted in gas and dust.

If the 10µm complex is not heated primarily by luminosity sources embedded within the IRS objects, it is interesting to consider how the dust in the central parsec might be heated to temperatures exceeding 250K. Becklin, Gatley, and Werner (1982) established on the basis of far-infrared observations that some $1 - 3 \times 10^7$ $L_\odot$ are emitted in the galactic nucleus, and Henry, DePoy, and Becklin (1984) demonstrated that nearly all of this luminosity may be attributable to IRS 16/Sgr A*. Following Scoville and Kwan (1976), we calculate that grain temperatures in excess of 300K can be obtained assuming a single source of 2×10^7 $L_\odot$ at the center of a distribution of grains 2 pc

from the source. Thus, it appears that the temperature structure
deduced from our new 10 μm images is consistent with the presence of a
central engine at the galactic nucleus.

The principal conclusions of this work are: 1) The "northern arm"
and "ionized bar" are not conspicuous color temperature features.
These large-scale structures are nearly constant in temperature and
cooler than the material surrounding them. IRS 1, 5 and 10 are
slightly warmer than the surrounding northern arm structure, and could
be partially self-luminous. However, IRS 2, 4, 6, 9, and 21 are not
distinguishable in color temperature from the surrounding material in
the bar structure. 2) The color temperature distribution suggests
that the infrared source complex is heated externally, possibly by a
dominant central luminosity source. 3) The total dust mass density we
derive in the central parsec is much smaller than the density in the
2 pc radius ring encircling the nucleus, and IRS 16/Sgr A* was not
detected in these observations, further evidence for the lower density
of interstellar material immediately surrounding the Galactic Center.

Considered in the context of previous observations, the derived
dust opacity and color temperature distributions are consistent with
the predictions of "central engine" models for the energetics of the
Galactic Center. These results do not determine the nature of this
luminosity source, but they do indicate that internal heating is not
apparent, and that a significant fraction of the material in the
complex is externally heated. Successful models for the energetics of
the inner few parsecs should, therefore, address a central source of
luminosity at the Galactic Center.

Aitken, D. K., Allen, M. C., and Roche, P. F. 1982, in The Galactic
  Center, ed. G. R. Riegler and R. D. Blandford, AIP Conference
  Proceedings No. 83, p. 67.
Becklin, E. E., Gatley, I., and Werner, M. W. 1982, Ap. J., 258, 135.
Becklin, E. E., Matthews, K., Neugebauer, G. and Willner, S. P. 1978a,
  Ap. J., 219, 121.
Brown, R. L. and Liszt, H. 1984, Ann. Rev. Astr. Ap., 22, 223.
Gatley, I. and Becklin, E. E. 1981, in Infrared Astronomy, IAU
  Symposium 96, ed. C. G. Wynn-Williams and D. P. Cruikshank,
  Dordrecht: D. Riedel Publ. Co., p. 281.
Genzel, R., Watson, D. M., Crawford, M. K., and Townes, C. H. 1985,
  Ap.J., in press.
Gezari, D. Y., Tresch-Fienberg, R., Fazio, G. G., Hoffmann, W. F.,
  Gatley, I., Lamb, G., Shu, P., and McCreight, C. 1985, Ap. J.,
  299, 1007.
Harris, A. I., Jaffe, D. T., Silber, M., and Genzel, R. 1985, Ap. J.
  (Letters), 249, L93.
Henry, J. P., DePoy, D. L., and Becklin, E. E. 1984, Ap. J. (Letters),
  285, L27.
Lacy, J. H., Townes, C. H., Geballe, T. R., and Hollenbach, D. J.
  1980, Ap. J., 241, 132.

Lamb, G. M., Gezari, D. Y., Shu, P., Tresch-Fienberg, R., Fazio, G. G., Hoffmann, W. F., and McCreight, C. R., 1984, Proc. S.P.I.E., 445, 113.
Lebofsky, M. J., Rieke, G. H., Deshpande, M. R., and Kemp, J. C. 1982, Ap. J., 263, 672.
Lo, K.-Y., and Claussen, M. J., 1983, Nature, 306, 647.
McCreight, C. R. and Goebel, J. H. 1981, Appl. Optics, 20, 3189.
McKelvey, M. E., McCreight, C. R., Goebel, J. H., Reeves, A. A. 1985, NASA TM #86667 (submitted to Appl. Optics).
Parry, C. M. 1980, Proc. S.P.I.E., 244, 2.
Parry, C. M. 1983, in Proc. Infrared Detector Technology Workshop, NASA Ames Research Center, ed. C. R. McCreight.
Rieke, G. H. and Lebofsky, M. J. 1982, in The Galactic Center, ed. G. R. Riegler and R. D. Blandford, AIP Conference Proceedings No. 83, p. 194.
Rieke, G. H., Telesco, C. M., and Harper, D. A. 1978, Ap. J., 220, 556.
Serabyn, E. and Lacy, J. H. 1985, Ap. J., in press.

BRACKETT ALPHA IMAGES OF THE GALACTIC CENTER

W. J. Forrest, M. A. Shure, J. L. Pipher, and C. E. Woodward
University of Rochester, Rochester, N.Y. 14627

ABSTRACT

Images of the central parsec of the galaxy have been obtained in the Brα emission line of hydrogen. The images are, for the most part, very similar to 5 and 15 GHz radio continuum images of this region. Alignment of the Brα with the radio images gives an improved position of the Sgr A* radio point source with respect to the infrared sources: 5.50" S and 0.15" W of IRS 7, a position remarkably devoid of compact objects in the near infrared. This position rules out identification of Sgr A* with IRS 16 Centre or IRS 16NW. A number of compact Brα sources with no radio counter- parts are seen. There is a string of four sources just north of Sgr A*, which itself is located in a Brα void. An isolated Brα source was also seen ≈ 10" SW of Sgr A*. These Brα sources may arise in very dense regions (optically thick in the radio), may be very cold, or may result from collisional excitation. They may also indicate the presence of local exciting sources in the central parsec.

OBSERVATIONS

The University of Rochester's 32x32 InSb array camera (Forrest et al 1985) was mounted on the NASA IRTF 3m telescope to image the central parsec of our Galaxy in the Brα (n= 5 to 4) emission line of hydrogen. The line emission was separated from the continuum with a 1.3% spectral resolution CVF. The pixels were 0.42" square giving a 13" array field of view. 88 images of the central 23" region were obtained during one hour on 22 August 1986 UT. Images were obtained sequentially at 4.10, 4.052, and 4.00 μm at 10 different telescope positions. Brα images (figs. 1 and 2) were derived by subtracting the continuum emission, at 4.1 and 4.0 μm, from the 4.052 μm image which is centered on the line. The two continuum images were scaled to eliminate (down to the 1% level) the bright continuum sources IRS 7 and 3. The surface brightness, calibrated with standard stars, is accurate to 20%. The point spread function, as gauged by the IRS 7 and 3 point sources, is 1" FWHM. The RMS noise in the Brα image was judged, from the fluctuations seen to the SW of IRS 7, to be one fifth of the lowest contour shown in fig. 1. The dynamic range of the image is 40:1.

DISCUSSION

Many of the features seen in the Brα images (figs. 1 and 2) can be traced also in the 5 and 15 GHz images (fig. 3, Yusef-Zadeh, personal communication; Lo and Clausen 1983; Lo 1986) which, aside from Sgr A*, are dominated by free-free emission from thermal ionized hydrogen and helium. In particular, both show bright regions near IRS 1W and IRS 13 and 2, the "northern arm" curving upward from IRS 16 through IRS 1W and 10, and continuing northward, and the "bar" running through IRS 13 and 2 and continuing to the SE past IRS 9 at the lower left. These same features are also seen in the Brγ map of Storey and Allen (1983, hereafter SA) with the exception of the bar structure. This similarity indicates these aspects of the Brα emission may be explained as recombination radiation from ionized hydrogen, similar to that occuring in thermal HII regions and planetary nebulae.

POSITION OF SGR A*

Alignment of these common features on the radio and Brα maps leads to a very accurate position for the Sgr A* point source: 5.50" S and 0.15" W of the bright point-like source IRS 7. The 5 GHz, 1" resolution maps of Lo and Clausen (1983) and Yusef-Zadeh (1986 personal communication, fig. 3) and a 0.66x0.34" resolution 15 GHz map of Yusef-Zadeh (1986, personal comm.) were used for the alignment; the three determinations all agreed to within 0.05". The position of IRS 7 was determined from the infrared images before subtraction of the continuum (not shown here). The Sgr A* position is 0.3" S of the position derived by Forrest et al (1986, hereafter FPS) from astrometry, in a region remarkably devoid of compact sources in the near infrared. In particular, this new position rules out identification of Sgr A* with the blue sources IRS 16 Centre (SA) or IRS 16NW (FPS, Stein and Forrest 1986). Using the offset derived above and the position of Sgr A* given by Brown et al. (1981) results in a position for IRS 7 in excellent agreement (<0.1" difference) with the astrometric position of IRS 7 found by FPS.

UNUSUAL BRACKETT ALPHA SOURCES

Of considerable interest are a number of compact Brα sources which have no counterparts in the 5 or 15 GHz radio maps mentioned above. This indicates unusual excitation conditions in these regions as discussed below. Just north of Sgr A*, adjacent to the "northern arm" and "bar" regions is a string of four such Brα sources (crosses in fig. 2). Each of these Brα sources is quite near (within 0.3") compact sources seen in the 2 μm continuum images of this region (SA and FPS). From the east, the 2 μm sources are IRS 16NE, 16 Centre, 16NW, and 1.6" W of IRS 16NW. The brightest of these Brα sources, near IRS 16NE, was also seen in Brγ by SA. Because of its proximity, this source of Brackett emission lines may be excited by the blue source IRS 16NE, as suggested by SA. The other Brα sources here may also be physically associated with their 2 μm counterparts. It is significant that these sources appear slightly elongated in Brα (figs. 1 and 2), which would be inconsistent with a stellar, point-like source. The final unusual Brα source is located ≈ 10" SW of Sgr A*, in a region devoid of 5 and 15 GHz radio emission and named infrared sources. This source is notable for its essentially stellar nature and quite weak 4 μm continuum flux - the line to continuum flux ratio is 1.4 for this source versus 0.4 for the IRS 16 NE, Centre and NW sources and 0.1 for the source 1.6" W of IRS 16NW.

If the Brα emission is due to hydrogen recombination, the relative lack of 5 and 15 GHz emission could be due to an extremely low electron temperature or a large free-free optical depth. The upper limits of 5 and 15 GHz flux to Brα flux are 2 to 4 times less than the bulk of the region shown in figs. 1 and 2. Since this ratio scales as $T_e^{0.85}$ for optically thin regions under "Case B" recombination conditions, this would imply an electron temperature at least a factor of 2 to 5 times lower in these sources than their surroundings. This would be an extremely low and unusual temperature. Alternatively, if the regions are very small and dense, the radio emission could be suppressed through self-absorption. Even for a low electron temperature of 3000K, the absence of 15 GHz emission requires that the sources be <0.1" in diameter. This appears to conflict with the extended nature of the four sources north of Sgr A*. Further, such a small, dense source would tend to expand on time scales of only several hundred years.

These weak radio, strong Brα sources are reminiscent of stars losing mass through a dense ionized wind. Examples are the young stellar objects, Be stars and Wolf-Rayet stars, though the latter are hydrogen deficient. It would be interesting if a large concentration of this type of object were at the galactic center. The 2 μm sources IRS 16 NE, Centre, and NW are all relatively blue with near infrared colors consistent with hot stars, strengthening the argument for this type of association.

Sternberg (1986) has suggested that a possible signature of the source responsible for the 511 keV positron annihilation line is strong and variable Brackett emission lines. The Brackett lines can be excited by collision and promptly decay, whereas radio free-free would be limited by the recombination time, typically 1 to 10 years. It is somewhat disturbing that there are 5 unusual sources, however. We plan to monitor the Brα emission from this region for variablility. In any case, the collisional excitation mechanism is a possible way of generating strong Brackett lines without accompanying radio free-free emission.

Finally, the conditions causing these unusual Brackett α sources could be related to the conditions found in active galactic nuclei. If so, it would be very profitable to understand these sources in great detail.

As noted above, in all cases the unusual Brα sources in the central parsec indicate local unusual conditions are present at their locations. This suggests the existence of local exciting sources scattered throughout this region.

This work was supported by grants from the NSF, NASA Ames, and the National Geographic Society.

REFERENCES

Brown, R.L., Johnston, K.J. and Lo, K.Y. 1981 Ap.J., 250, 155.
Forrest, W.J., Moneti, A., Woodward, C.E., Pipher, J.L., and Hoffman,

A. 1985, Pub. A.S.P., 97, 183.
Forrest, W.J., Pipher, J.L. and Stein, W.A. 1986, Ap. J. (Letters), 310, L49 (FPS).
Lo, K.Y. and Claussen, M.J. 1983, Nature, 306, 647.
Lo, K.Y. 1986 Science, 233, 1394.
Stein, W.A. and Forrest, W.J. 1986 Nature, 323, 232.
Sternberg, A. 1986 Ap.J., 301, 923.
Storey, J.W.V. and Allen, D.A. 1983, M.N.R.A.S., 204, 1153 (SA).

Figure 1. (below) The Brα image of the central parsec with 40:1 dynamic range and 1" resolution. Crosses mark the positions of IRS 7 and 3 and tick marks indicate the Sgr A* position derived here. The greybar on the left shows the intensity scaling employed.

Figure 3. (below) A 5 GHz radio map of the central parsec with 1" resolution courtesy of Farhad Yusef-Zadeh (1986, personnal comm.). The first 12 contours are 1 to 12 times 5 mJy/beam.

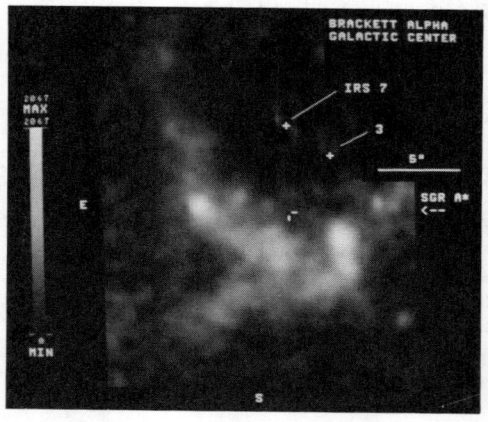

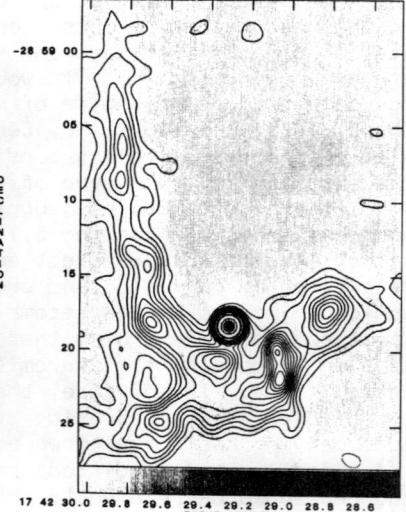

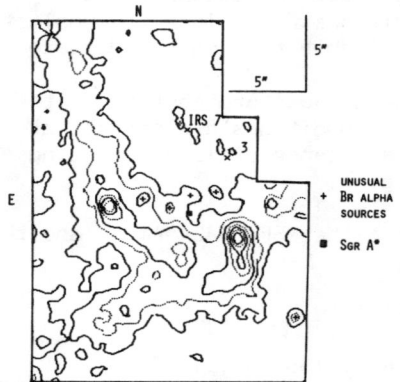

Figure 2. (left) The Brα contour map. Contours are 1 to 7 times 0.016 ergs/cm^{2}sec ster, with even contours dotted. The positions of five Brα sources without radio counterparts are marked with crosses (+). A solid square indicates the Sgr A* position derived here and X's mark the positions of IRS 7 and 3.

PRELIMINARY RESULTS OF THE SPACELAB-2 INFRARED TELESCOPE
SURVEY OF THE GALACTIC PLANE AT $2.4\,\mu m$

G. J. Melnick, G. G. Fazio, D. G. Koch
Smithsonian Astrophysical Observatory

G. H. Rieke, E. T. Young, F. J. Low, W. F. Hoffmann
Steward Observatory, University of Arizona

T. N. Gautier
Jet Propulsion Laboratory, California Institute of Technology

ABSTRACT

We present preliminary $2.4\,\mu m$ maps of the galactic center region and the galactic plane between $\ell = -10°$ and $110°$ obtained with the Infrared Telescope (IRT) flown aboard the Space Shuttle in 1985. These results are in qualitative agreement with earlier balloon-borne observations which showed the overall $2.4\,\mu m$ emission to be more widely distributed perpendicular to the plane than is seen in the longer wavelength IRAS survey. We measure the FWHM of the surface brightness perpendicular to the plane to be $\sim 5°$ for $15° \leq \ell \leq 60°$ and somewhat wider inward of $15°$ owing to the presence of an extended bulge. The intensity of the surface brightness along the plane shows a steady increase toward the galactic center for $\ell < 70°$, with a sharp increase seen within $\pm 2°$ of the galactic center. The structure within the brightness distribution along the plane is tentatively associated with various spiral arms and HII regions.

INTRODUCTION

The $2.4\,\mu m$ radiation we detect is predominantly emitted by late-type stars which constitute the major fraction of the mass in the Galaxy. This component therefore governs galactic dynamics. Conversely, longer wavelength emission ($\lambda \geq 10\,\mu m$) samples mainly the thermal emission from dust grains heated by the absorption of UV radiation from early-type stars. This component is largely associated with active star forming regions. A comparison of the $2.4\,\mu m$ surface brightness with the longer wavelength surface brightness can therefore be used to better understand the relationship between the large-scale mass distribution in the disk and the smaller regions of the plane which experience enhanced star formation.

Partial surveys of the galactic plane at $2.4\,\mu m$ have been made by several groups (e.g., refs. 1,2,3,4, and refs. therein). The primary advantages of the present survey are: (1) greater sensitivity - the IRT is more than 10 times as sensitive as previous balloon-borne $2.4\,\mu m$ surveys, (2) greater spatial coverage - the IRT was able to survey approximately 75% of the plane during the mission, and (3) better spatial registration - the positional errors of the IRT are estimated to be 5' versus ~ 12' for the balloon[1]. The final calibration of the IRT is not yet complete, making a quantitative presentation of absolute intensity maps inappropriate at this time. Nevertheless, a number of general conclusions can be drawn from an examination of the relative surface brightness maps now available.

OBSERVATIONS

The IRT was flown between July 29 and August 6, 1985, aboard the Space Shuttle as part of the Spacelab-2 mission (51-F). The IRT was designed to be a large beam ($0.6° \times 1.0°$/detector), short duration (~ 10 days) instrument capable of surveying a large fraction of the sky from the Shuttle cargo bay during a 7-day mission. Unlike IRAS, which integrated the dewar and telescope into a single structure, the IRT consists of a separate dewar and telescope assembly. In this configuration, a stationary 250-liter superfluid helium dewar is connected to a movable cryostat containing the optics and detectors. The telescope, an f/4, 15.2-cm highly baffled Herschelian system, was able to scan $\pm 45°$ in a plane perpendicular to the payload bay long axis and centered on the payload bay vertical. Cooling to the telescope was provided by the boiloff from the superfluid helium, which was phase separated by a porous plug and carried to the focal plane, optics, and baffles via a helical coil through a rotary vacuum joint. The focal plane consisted of 10 discrete photoconductors covering the wavelength range between 1.7 and 118 μm in six broad passbands plus a cold shutter. Both the focal plane and shutter were maintained at 3.1 K throughout the flight, while the optics and lower baffles were kept at ≤ 8 K. A more complete description of the instrument can be found in Koch et al.[5] (and refs. therein).

The 2.4 μm detector had a FWHM bandpass of 1.3 μm (1.7 - 3.0 μm). In order to be able to obtain two-dimensional positions from known point sources, an N-slit mask was placed immediately in front of this detector resulting in a field of view of 3 times 5.6 by 1° ($\simeq 8.5 \times 10^{-5}$ steradian). All of the data presented here were obtained with the telescope scanning and the Shuttle pitching to provide mapping of 60% of the sky during each orbit (versus a point and integrate mode). Absolute sky positions were derived from the Orbiter inertial navigation data plus an accurate knowledge of the telescope scan angle. These computed positions were verified frequently by comparison with the known positions of bright 2 μm point sources. The total sky coverage is shown in Figure 1.

RESULTS

Preliminary maps of the galactic center region and the first quadrant of the galactic plane are shown in Figures 2 and 3. Our results are in qualitative agreement with the earlier maps from the balloon surveys, i.e., we confirm the presence of the two strong emission peaks which straddle the galactic center at $\ell = -0.3°$, $b = +0.5°$, and $\ell = 0.0°$, $b = -0.9°$. The position of these peaks agrees with those shown in the map produced by Oda et al.[4], but are slightly different from the positions shown in the map produced by Matsumoto et al.[6], who place the northern peak at the center ($\ell = 0.0°$, $b = 0.0°$). Our first quadrant map is in qualitative agreement with the maps

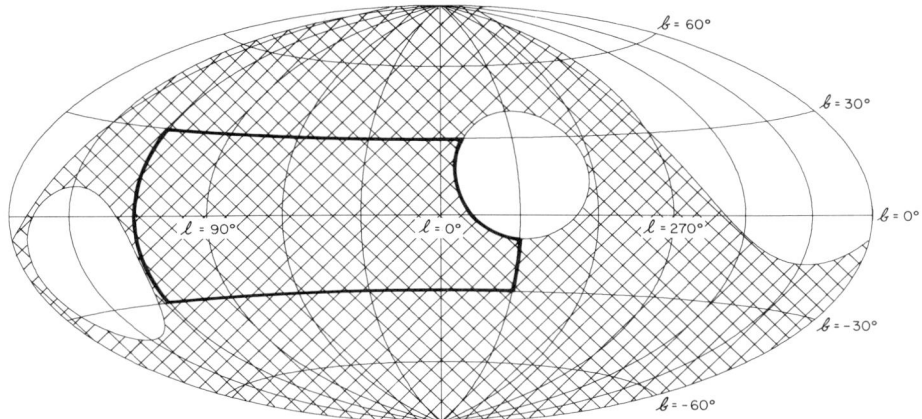

Fig. 1. The cross-hatched region represents the total sky coverage at 2.4μm. The area outlined with the dark box is the region for which the data has been processed to date.

presented by Hayakawa et al.[1] and Oda et al.[4]

The most prominent features of the maps produced to date are: (1) the width of the 2.4μm emission perpendicular to the galactic plane, and (2) the structure in surface brightness profile along the plane.

The FWHM of the 2.4μm emission perpendicular to the plane for a range of longitudes is shown in Figure 4. Several things are clear. For $15 \le \ell \le 60°$ the average FWHM of the 2.4μm surface brightness is

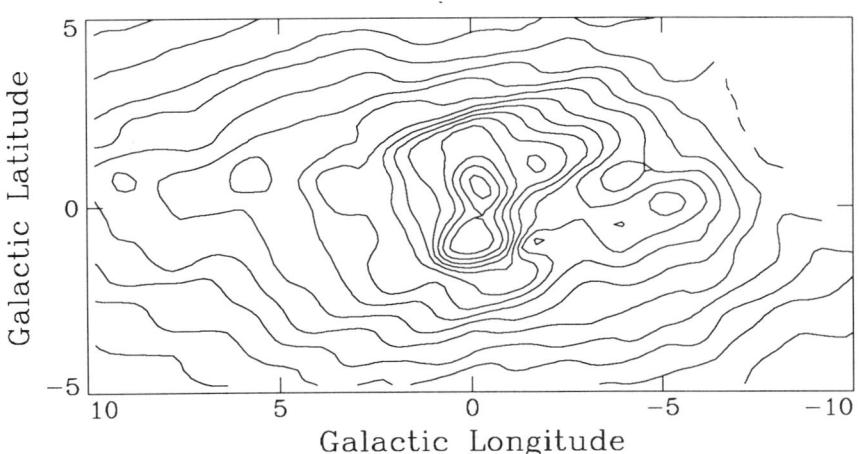

Fig. 2. The map of 2.4μm surface brightness of the galactic center. The absolute intensity scale has not yet been established.

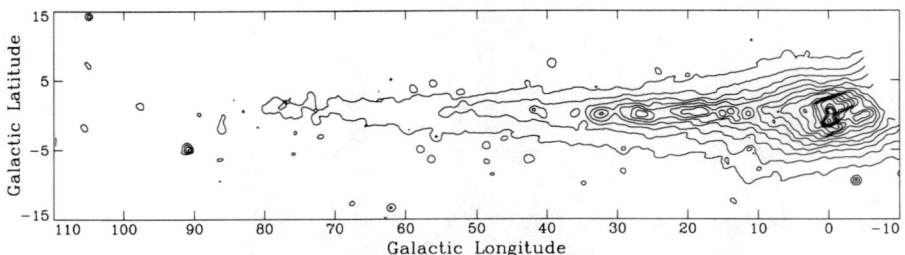

Fig. 3. The map of 2.4 μm surface brightness of the first quadrant of the galactic plane. The absolute intensity scale has not yet been established.

approximately 5°. For longitudes greater than 60°, the increase in the 2.4 μm surface brightness at the plane is too small to assign an accurate FWHM value. Inward of $\ell \simeq 15°$ the effect of the nuclear bulge is evident in the increased width of the 2.4 μm surface brightness. The apparent decrease in the FWHM at $\ell = 0°$ is the result of a sharp increase in the 2.4 μm surface brightness within ± 2° of the center (see Figure 5) rather than a decrease in the bulge component at this longitude. These results contrast with the results obtained by IRAS: within ± 6° of the galactic center, the FWHM of the 12 to 100 μm data is $\leq 1°$,[7] and from $55° \leq \ell \leq 305°$ the 60 μm is consistent with a FWHM of $< 2°$.[8]

A profile of the 2.4 μm surface brightness along the plane (b = 0.0°) is shown in Figure 5. Interpreting this plot without corresponding information about the extinction along the plane is

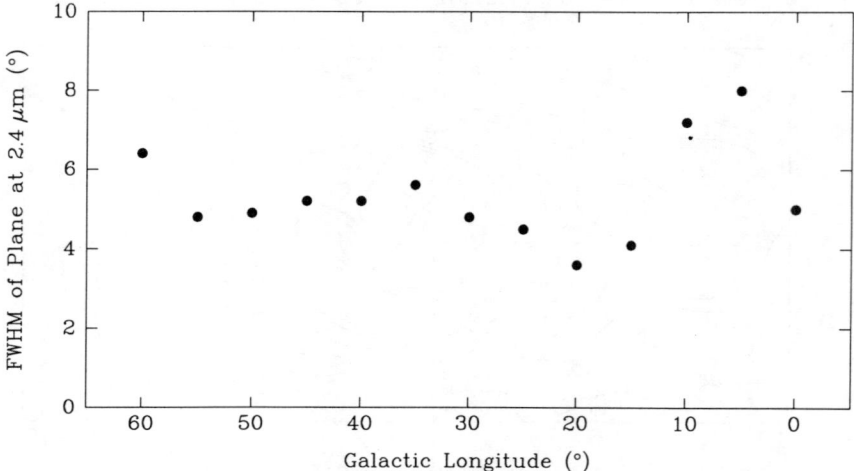

Fig. 4. The FWHM of the 2.4 μm surface brightness perpendicular to the galactic plane as a function of galactic lonitude.

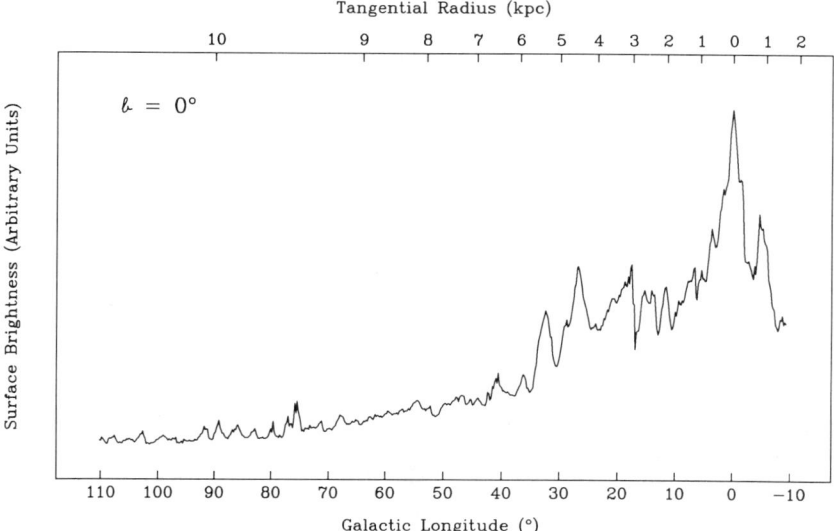

Fig. 5. Longitudinal distribution of 2.4μm surface brightness.

difficult. However, it is tempting to associate many of the secondary maxima with well known H II regions or spiral arms: NGC 6334 ($\ell \sim -9°$), W28 ($\ell \sim 6°$), M17 ($\ell \sim 15°$), M16 ($\ell \sim 16°$), the 3 kpc arm ($\ell \sim 18°$), the 5 kpc arm ($\ell \sim 27°$), the tangent to the Scutum arm ($\ell \sim 33°$), and the tangent to the Cygnus arm ($\ell \sim 75°$). As noted by Hayakawa et al.[1], H II complexes relatively near the Sun at $\ell \sim 7°$ and 17° may also contribute.

REFERENCES

1. S. Hayakawa, T. Matsumoto, H. Murakami, K. Uyama, J.A. Thomas, and T. Yamagami, Astron. & Astrophys., 100, 116 (1981).
2. S. Hayakawa, T. Matsumoto, H. Murakami, K. Uyama, T. Yamagami, and J.A. Thomas, Nature, 279, 510 (1979).
3. W. Hoffmann, D. Lemke, and A. Frey, Astr. Ap., 70, 427 (1978).
4. N. Oda, T. Maihara, T. Sugiyama, and H. Okuda, Astr. Ap., 72, 309 (1979).
5. D. Koch, G. G. Fazio, W. A. Traub, G. H. Rieke, T. N. Gautier, W. F. Hoffmann, F. J. Low, W. Poteet, E. T. Young, E. W. Urban, and L. Katz, Optical Eng., 21, 141 (1982).
6. T. Matsumoto, S. Hayakawa, H. Koizumi, H. Murakami, K. Uyama, T. Yamagami, and J. A. Thomas, The Galactic Center, AIP Conference Proceedings No. 83, eds. G. R. Riegler and R. D. Blandford, p. 48 (1982).
7. T. N. Gautier, M. G. Hauser, C. A. Beichman, F. J. Low, G. Neugebauer, H. H. Aumann, N. Boggess, J. P. Emerson, S. Harris, J. R. Houck, R. E. Jennings, and P. L. Marsden, Astrophys. J., 278, L57 (1984).
8. N. Scoville and J. C. Good, IRAS preprint (1986).

ASTROMETRIC POSITION OF IRS-7 IN THE GALACTIC CENTER

E.E. Becklin
University of Hawaii

H. Dinerstein*
University of Texas

I. Gatley*
UKIRT

M.W. Werner*
NASA-Ames

B. Jones
Lick Observatory

We have used the offset pointing capabilities of the Infrared Telescope Facility together with new astrometric positions of seven stars within 7 arcmin of the galactic center which are bright at both infrared and optical wavelengths to determine an absolute position for IRS-7. The positions of the seven stars were determined from two plates of epoch 1952 and 1982, relative to 16 "Perth 70" stars (Hog and Heide, 1976). The position of IRS-7 relative to these seven stars was determined by repeatedly offsetting between the stars and IRS-7 and carefully centering each of the objects in a 3.8 arcsec beam at 1.65 μm. The resultant 1950 position of IRS-7 is:

$$RA = 17h\ 42m\ 29.32s\ +/-\ 0.02s$$

$$DEC = -28°\ 59'\ 13.0'' +/- 0.2''$$

This position agrees well with those determined independently by Storey and Allen (1983) and Forrest et al. (1986), a value of 1.4"E and 5.5"S is found for this offset. Taking this result and the position of the non-thermal radio source Sgr A* from Brown et al. (1981), we find that Sgr A* is 1.3" +/- 0.3" W and 0.1" +/- 0.3" S of IRS-16 center.

REFERENCES

R.L. Brown, K.J. Johnston, K.Y. Lo, *Ap. J.* **250**, 155 (1981).
W. Forrest, J. Pipher, and W. Stein, *Ap. J. Letters* **301**, L49 (1986).
E. Hog and J. von der Heide, *Abhandlungen der Hamburger Sternwarte*, Band IX.
J.W.V. Storey and D.A. Allen, *MNRAS* **204**, 1153 (1983).
E. Tollestrup, R.W. Capps, and E.E. Becklin, *BAAS* **18**, 640 (1986).

*Visiting Astronomer at the Infrared Telescope Facility Operated by the University of Hawaii under contract to NASA.

PROPER MOTION OF THE COMPACT, NONTHERMAL RADIO SOURCE IN THE GALACTIC CENTER

D. C. Backer
Radio Astronomy Laboratory and Astronomy Department
University of California, Berkeley, 94720.

R. A. Sramek
National Radio Astronomy Observatory
Socorro, NM, 87801.

ABSTRACT

Observations with the VLA for four epochs from 1981 to 1985 have resulted in the first two-axis measurement of the proper motion of the compact, nonthermal radio source in the galactic center, SgrA*. This result considerably improves our past single-axis measurement (Backer and Sramek 1982). The observed motion is consistent with that expected for an object in the galactic center whose peculiar velocity is no larger than 40 r km s^{-1}, where r is the distance to the galactic center in units of 8.5 kpc. If there is a massive black hole in the galactic center, then these observations suggest that the radio emission must be coming from a synchrotron corona surrounding the hole. The converse is not true: these observations, by themselves and at their present level of accuracy, do not require the existence of a massive black hole.

INTRODUCTION

The expected motion for an object at rest in the center is a combination of both the solar peculiar motion with respect to the local standard of rest expressed as an angular velocity about the galactic center, and the galactic rotation rate at the distance of the sun given by Oort's constants, A-B. The latter leads to a pure longitude motion that may be computed from either the 1976 IAU standard constants (Allen 1973) of −250 km s^{-1}/10 kpc (−5.3 mas yr^{-1}), or from the 1985 IAU values (Kerr and Lynden-Bell 1985) of −220 km s^{-1}/8.5 kpc (−5.5 mas yr^{-1}). The solar motion results in an additional apparent motion of −0.32 or −0.38 mas yr^{-1} in longitude, and −0.15 or −0.18 mas yr^{-1} in latitude, for galactic center distances of 10 kpc and 8.5 kpc, respectively. The total expected motion is then

$$\mu_l = -5.60 \quad \text{or} \quad -5.88 \quad \text{mas yr}^{-1}$$

$$\mu_b = -0.15 \quad \text{or} \quad -0.18 \quad \text{mas yr}^{-1}$$

EXPERIMENT

In our differential astrometry experiment at the VLA we measure the position of SgrA* with respect to three reference sources. The reference sources are spaced in a triangle around SgrA* at an average distance of 0.5°. Interferometer phases from each observation are first calibrated using 1748-253, and then combined to produce an instantaneous position offset. The offsets for both SgrA* and the reference sources are dominated by a common temporal variation of 0.05″ and 0.15″ in RA and DEC, respectively. These variations are the result of tropospheric refraction in the skies above the Plains of St. Augustin. The variations are large owing to the low elevation of the galactic center.

Further analysis of the position offsets leads to a single differential position for SgrA* on each day of each epoch. The reference phases are first interpolated to the times of observation of SgrA*, and then subtracted from the SgrA* values. The resultant differential phases are fitted to a net position offset. These offsets are displayed in Figure 1 along with the proper motion fit. Results from only one day, in epoch 3, have been excluded from this plot; on that day tropospheric phase fluctuations were so extreme that phases could not be tracked from one ten-minute observation of SgrA* to the next on the critical long baselines. Our preliminary result from these data is

$$\mu_\alpha = -3.55 \pm 0.20 \quad \text{mas yr}^{-1}$$

$$\mu_\delta = -4.83 \pm 0.80 \quad \text{mas yr}^{-1}$$

These can be transformed to galactic coordinates to yield

$$\mu_l = -5.95 \pm 0.70 \quad \text{mas yr}^{-1}$$

$$\mu_b = +0.43 \pm 0.50 \quad \text{mas yr}^{-1}$$

The errors are rough estimates of the uncertainty derived by processing the data by slightly altered techniques. Analysis of the data using only pairs of reference sources shows consistent results, and indicates that all reference sources are extragalactic.

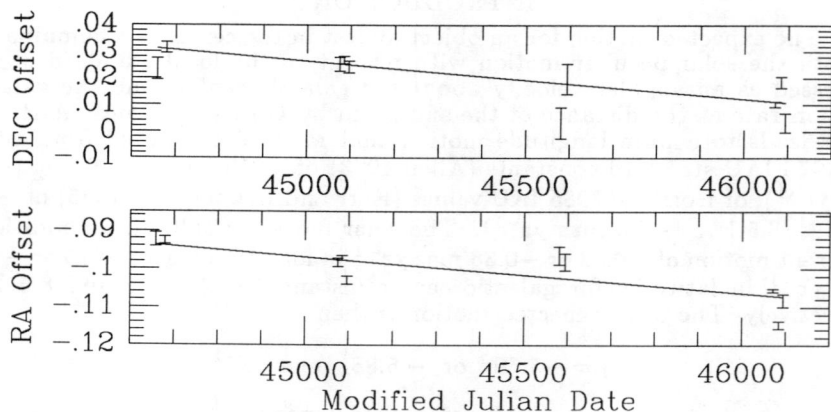

Fig. 1 Position offsets for SgrA* from VLA observations in four epochs from 1981 to 1985. Results from independent days at each epoch are displaced for presentation. Solid line is fit for proper motion discussed in text.

DISCUSSION

The residual between the measured proper motion and the expected motion discussed initially is

$$\mu_l = -0.35 \quad \text{or} - 0.07 \pm 0.70 \quad \text{mas yr}^{-1}$$

$$\mu_b = +0.58 \text{ or } +0.61 \pm 0.50 \text{ mas yr}^{-1}$$

Neither of these are significant with respect to zero. A tighter comparison can be made using the right ascension motion. The residual motion in RA is -0.90 or -0.80±0.20 mas yr^{-1} which is barely significant in comparison to zero if our error estimate is one standard deviation. This comparison uses the galactic constants, and is therefore less certain as a statement about the peculiar motion of SgrA*.

We use twice the limit on the latitude motion, 1.0 mas yr^{-1}, to place an upper bound of 40 r km s^{-1} on the peculiar velocity of SgrA*. This limit probably applies in both coordinates. If there is a five-million solar mass massive black hole in the galactic center (Crawford et al. 1985), then the small transverse velocity of SgrA* makes a unique identification of SgrA* with a radio-emitting corona around the hole. If one does not accept the evidence of Crawford et al., then our proper motion determination must be pushed to lower limits to provide evidence that its peculiar motion is much different than that of the ionized gas in the center (Serabyn and Lacy 1985).

This research has been supported at the Radio Astronomy Laboratory by NSF grant AST-81147147. The National Radio Astronomy Observatory is operated by Associated Universities, Inc., under contract with the National Science Foundation.

REFERENCES

C. W. Allen, *Astrophysical Quantities, Third Edition* (The Athlone Press, London, 1973).

D. C. Backer and R. A. Sramek, *Ap. J.* **260**, 512 (1982).

M. K. Crawford, R. Genzel, A. I. Harris, D. T. Jaffe, J. H. Lacy, J. B. Lugten, E. Serabyn, and C. H. Townes, *Nature* **315**, 467 (1985).

F. J. Kerr, and D. Lynden-Bell, *Highlights of Astronomy* **7**, 00 (1985).

E. Serabyn, and J. H. Lacy, *Ap. J.* **293**, 445 (1985).

THE DISTANCE TO THE CENTER OF THE GALAXY

J. M. Moran, M. J. Reid, M. H. Schneps, C. R. Gwinn
Harvard–Smithsonian Center for Astrophysics, Cambridge, MA 02138

R. Genzel
University of California, Berkeley, CA 94720

D. Downes
Institut de Radio Astronomie Millimetrique, Grenoble, France

B. Rönnäng
Onsala Space Observatory, Onsala, Sweden

ABSTRACT

We have estimated the distance to the center of the galaxy from the measurements of the proper motions of 24 H_2O maser spots in the source Sgr B2-North. A comparison of the transverse angular velocities measured from VLBI observations and the line of sight Dopper velocities gives a distance estimate of 7.1 ± 1.5 kpc.

The distance to a star forming region can be determined by measuring the proper motions within H_2O maser clusters. If the motions of the maser spots are random, the distance can be determined by applying the technique known as statistical parallax. Alternatively, if organized motions are evident in the proper motions, one can model the source to estimate its the distance. Both methods rely on a comparison of the radial component of the motion (in km/s) and the proper motion on the plane of the sky (in milli-arcseconds/year).

Velocity components toward the observer are determined from Doppler shifts, and angular motions across the plane of the sky are measured by VLBI imaging. Since, in a statistical sense, the transverse velocities should equal the line of sight velocities, the distance to the cluster can be determined by comparison of the angular and line-of-sight velocities. Results for the H_2O masers associated with Orion A[1] and with W51[2,3] H_2O masers have demonstrated the power of this technique.

We have monitored the motions of many H_2O maser clumps associated with star forming regions in the Galaxy with five VLBI observations from December 1980 through June 1982. In this paper we present results from the H_2O masers associated with the Sgr B2-North molecular cloud. Relative positional accuracies of about 10 μas have been achieved for H_2O masers across a 2 arcsec field, yielding relative transverse velocities with uncertainties of about 3 km/s over a 3 month time span. The proper motions of 24 features were measured reliably.

The proper motion information suggests that the Sgr B2-North masers are expanding. We have modeled the source as a uniformly expanding flow and estimated the parameters including the source distance, the expansion center, and the expansion velocity by fitting the kinematic data in a least-squares sense. The expansion center is near the densest concentration of H_2O features and the expansion velocity is approximately 40 km/s. This outflow is comparable in linear size but more energetic than the outflow associated with IRc 2 in Orion.

The distance to the Sgr B2-North H_2O masers is estimated to be 7.1 kpc with a formal uncertainty of $\pm$ 1.2 kpc. We are currently studying sources of systematic error. The total error is about $\pm$1.5 pc.

The following facts strongly suggest that Sgr B2-North is within a few hundred parsecs of the Galactic Center: 1) it is a unique cloud projected only 80 pc from the G.C.; 2) its LSR velocity of 64 km/s rules out a chance projection far from the Galactic Center since a velocity near zero would be expected in that case; 3) atomic and molecular absorption studies place Sgr B2-North inside the 270 pc expanding shell.[4] Therefore, our results for Sgr B2-North suggest a value for R_o considerably less than the 1966 IAU adopted value of 10 kpc and less than the 1985 value of 8.5 kpc.[5]

Previous studies by our group indicated distances to the Orion masers of 0.48 $\pm$ 0.08 kpc and to the W51 masers of 7 $\pm$ 1.5 kpc. Orion is too close to be used to determine R_o. The W51 result can be used to determine R_o, provided a kinematic model of the Galaxy is assumed. If Θ_o = 250 km/s and R_o = 7.1 kpc, W51's v_{LSR} = 57 km/s implies a distance of 5.5 kpc; for Θ_o = 220 km/s a distance close to 5 kpc is favored (although the LSR velocity is slightly above the tangent point value). Thus, our measured distance to W51 is about $1 - \sigma$ larger than its kinematic distance for R_o = 7.1 kpc. Put another way, the Schmidt rotation model with Θ_0 = 250 km/s and measured distance to W51 of 7 $\pm$ 1.5 kpc imply a value of R_o of 9 $\pm$ 3 kpc.

The following is a brief list of items which would be affected by reducing R_o: 1) decrease kinematic distances; 2) reduce the mass of the Galaxy and of the Galactic Center; 3) favor recent revisions of the absolute magnitudes of RR Lyrae variables which in turn affects distances in the Local Cluster, the extragalactic scale, and possibly H_o; 4) reduce the total luminosity of X-ray bursts, some of which appear super-Eddington for R_o = 10 kpc.[6]

Improvements in the distance to the Galactic Center will come from further analysis of our present data which includes Sgr B2-Middle and W49. In particular, W49 is fortuitously located at l^{II} = 43° with an LSR velocity near zero ($\approx$ 9 km/s); it is far from the tangent point and an estimate of R_o from its distance is not sensitive to details of the kinematic model of the Galaxy. W49 is a very strong source with hundreds of maser features and our data should yield a very precise distance. A detailed description of the results on Sgr B2-North will be published shortly.[7]

REFERENCES

1. R. Genzel, M. J. Reid, J. M. Moran, and D. Downes, *Ap. J.*, **244**, 884 (1981).
2. R. Genzel, D. Downes, M. H. Schneps, M. J. Reid, J. M. Moran, L. R. Kogan, V. I. Kostenko, L. I. Matveyenko, and B. Rönnäng, *Ap. J.*, **247**, 1039 (1981).
3. M. H. Schneps, A. P. Lane, D. Downes, J. M. Moran, R. Genzel, and M. J. Reid, *Ap. J.*, **249**, 124 (1981).
4. N. Z. Scoville, *Ap. J. (Letters)*, **175**, L127 (1972).
5. F. J. Kerr and D. Lynden-Bell, *M.N.R.A.S.*, **221**, 1023 (1986).
6. T. Ebisuzaki, T. Hanawa, and D. Sugimoto, *P.A.S. Japan*, **36**, 551 (1984).
7. M. J. Reid, M. H. Schneps, J. M. Moran, C. R. Gwinn, R. Genzel, D. Downes, and B. Rönnäng, *Ap. J.*, in preparation.

DON'T BE AFRAID TO OBSERVE THE NEXT SERIES OF LUNAR OCCULTATIONS OF THE GALACTIC CENTER, 1986 - 1989!

Aage Sandqvist
Stockholm Observatory
S-133 00 Saltsjöbaden, Sweden

ABSTRACT

A new series of lunar occultations of the Galactic Center takes place during the period 1986 - 1989. The previous series of Galactic Center occultations occurred in 1968 - 1970. That series revealed that the continuum radio source, Sgr A, consists of at least two components (now known as Sgr A West and Sgr A East) and that the continuum source is surrounded by a rotating and contracting cloud of dust and molecules. It could be rewarding to observe the new series of occultations at X-ray, infrared and radio wavelengths. Occultation predictions are available from the author upon request.

INTRODUCTION

During the next three years the Moon is going to treat us to a new series of lunar occultations of the Galactic Center. This is not a phenomenon that occurs very often - the last time was during the period 1968 - 1970. I wish to strongly encourage the astronomical community to take advantage of this occasion for studying the structure of the center of our Galaxy with milliarcsecond resolution.

The major disadvantage of lunar occultations is of course the low sensitivity of the technique resulting from the short integration times necessary to follow the motion of the Moon. This has led some astronomers to be wary of the method and ignore its power. But for strong sources, like Sgr A or IRS 16, the method offers resolution available only with radio interferometers. In the infrared wavelength region, this resolution may not become available again for decades. Lunar occultations are sensitive to extended, as well as fine, structure - in contrast to interferometers. The lunar ephemeris is accurate to 0.01 arcsec and the lunar limb is known to 0.1-arcsec accuracy. This means that absolute position determination is possible at different wavelengths to an accuracy of 0.1 arcsec (possibly greater with the use of lunar orbiter photographs) without worrying about different frames of references.

THE LUNAR OCCULTATIONS OF 1968

Some of the strength of the lunar occultation method can be illustrated with the results of the lunar occultations of the Galactic Center in 1968. They revealed that (i) Sgr A consists of at least two components, now known as Sgr A West and Sgr A East and (ii) the continuum source is surrounded by a rotating and contracting cloud of dust and molecules (Sandqvist 1971, 1974)[1,2]. Downes and Martin[3] demonstrated the thermal and nonthermal nature of the continuum

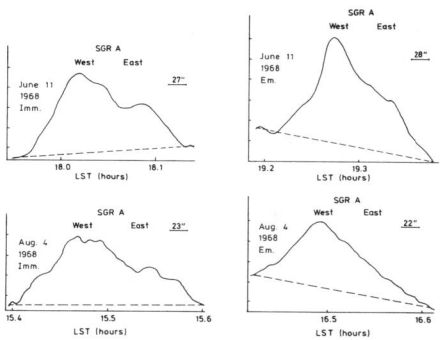

Figure 1. 18-cm strip brightness distributions from the 1968 occultations (Sandqvist 1971) [1].

components while using an interferometer in 1971, the infrared structure of the ring cloud was described in 1982 by Becklin, Gatley and Werner[4]. The kinematics of the ring have been further discussed by Lester et al.[5] in 1981, Genzel et al.[6] in 1982 and Liszt, Burton and van der Hulst[7] in 1985.

The restored strip brightness distributions of the Sgr A 18-cm continuum from two of the 1968 occultations[1] are reproduced in Figure 1. These curves give the structure across Sgr A at four different position angles and it is clear that Sgr A West and East had been detected. These data were combined with those from other 1968 occultations and a cleaned two-dimensional brightness distribution map[2] was produced (the "BEST" map, acronym for Beautiful Extraterrestrial Synthesis Telescope).

The 1665- and 1667-MHz OH and 1420-MHz HI maps[2], obtained with the 1968 occultations, are reproduced in Figure 2. They show the velocity-integrated (colummn density/excitation temperature)-distributions for an OH velocity range of -42 to +102 km s^{-1} and an HI velocity range of +27 to +73 km s^{-1}. Thes maps clearly show the northeast and southwest components of the Galactic Center ring structure (in addition to the dominant +40 km s^{-1} feature). The 1986 HCN map of Gusten et al.[8] is superimposed on the Figure 2 maps to facilitate a comparison of the detailed structure in the ring cloud as seen in the occultation maps and interferometer maps.

The velocity information, lost in the integrated occultation maps, is available in the original occultation data[1,9], with velocity resolutions of 2 to 6 km s^{-1} per channel. Figure 3, which is a

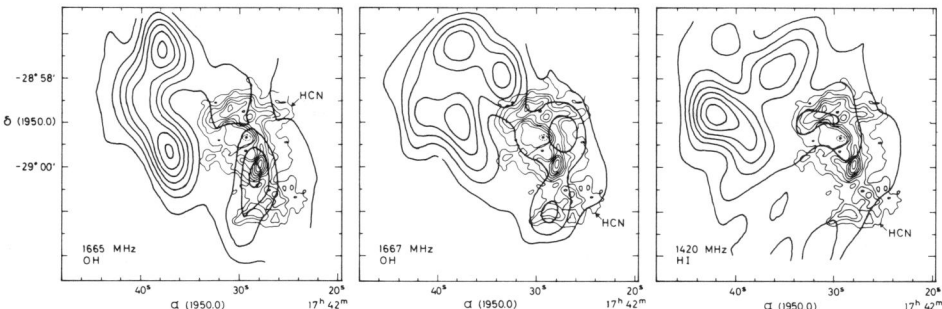

Figure 2. The velocity-integrated OH (-42 to +102 km s^{-1}) and HI (+27 to +73 km s^{-1}) contour maps obtained with the 1968 occultations (Sandqvist 1974)[2], and the HCN map of Gusten et al. (1986)[8].

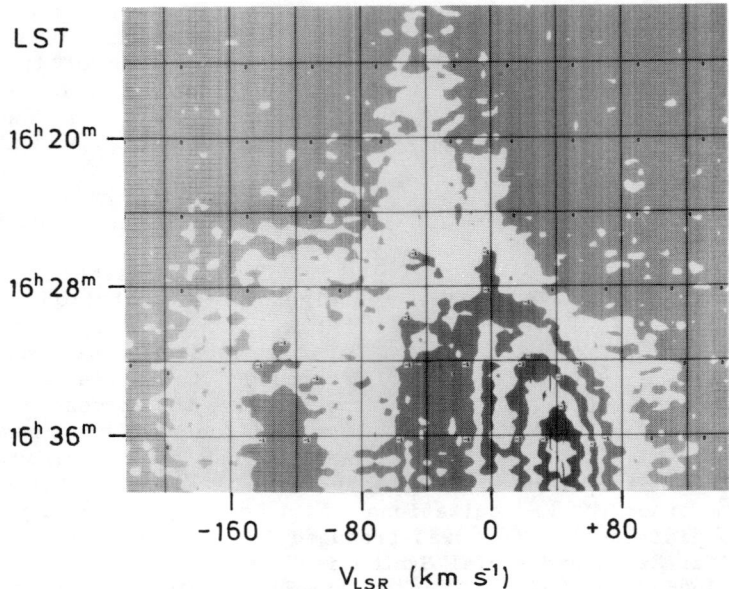

Figure 3. An occultation 1665-MHz OH line antenna temperature contour map (Sandqvist 1973)[8]. LST is local sidereal time, V_{LSR} is radial velocity, 1 contour unit = 7K.

reproduction of an occultation 1665-MHz OH line antenna temperature contour map[9], may serve as an example. The abcissa is radial velocity with respect to the local standard of rest, the ordinate is local sidereal time increasing downwards. During this emersion, the Moon gradually uncovered the molecular region, starting near the southwest ring component and moving northeastwards. The velocity centroid begins at negative velocities and gradually shifts towards positive velocities as the molecular region is uncovered. (Do not confuse with the -130 km s^{-1} feature). From this and similar maps, obtained at a number of different position angles, a model for the kinematics of the region was constructed.

Table I Atomic and molecular gas at the core of the Galaxy

	Species	radius (pc)	rotation velocity (km s^{-1})	infall velocity (km s^{-1})	Observations
Sandqvist (1974)[2]	OH,HI	10	50	50	1968 Lunar Occult.
Genzel et al. (1984)[10]	OI	6	70	?	NASA KAO
Liszt et al. (1985)[7]	HI,CO	3.2	110	50	VLA
Sandqvist et al. (1985)[11]	HCO$^+$	2.5	70	30-60	NRAO 12 m, Kitt Peak
Gatley et al. (1986)[12]	H$_2$	2	100	50	UKIRT
Serabyn et al. (1987)[13]	CO,CS	8	110	<20-30	IRAM 30 m

The model[2], proposed in 1974, of a rotating and contracting molecular dust cloud containing Sgr A, (radius 10 pc, rotation velocity 50 km s^{-1}, contraction velocity 50 km s^{-1}) does not differ too radically from the models being discussed today (see Table I).

OCCULTATION PREDICTIONS FOR THE 1986 - 1989 SERIES

Lunar occultations have definitely been beaten by today's interferometers at wavelengths greater than 3 mm. But in the sub-mm and infrared ranges it will probably be many years before lunar occultations' theoretical milliarcsecond resolution will be surpassed by other means. So, why not give it a try? We have run a number of occultation predictions for the upcoming Galactic Center series which are available upon request. The series began in March of 1986 and will continue until September of 1989. It is predominantly a northern hemisphere phenomenon with observatories at Cerro Tololo, La Silla, Parkes etc. being left out in the cold. Some indication of the occultation quality of several sites is given in the following list, where the first number in the bracket is the number of "good" occultations and the second number is typical elevation in degrees: Ootacamund (10,40), Mauna Kea (7,35), Palomar (6,20), Kitt Peak (5,25), VLA (5,20), Green Bank (5,20), La Palma (3,30), Pico Veleta (2,20), Nobeyama (2,20).

We shall be happy to supply predictions if you send us the longitude, latitude and altitude of your observatory.

REFERENCES

1. Aa. Sandqvist, Ph.D. Thesis (University of Maryland, 1971).
2. Aa. Sandqvist, Astron. Astrophys. **33**, 413 (1974).
3. D. Downes and A. H. M. Martin, Nature **233**, 112 (1971).
4. E. E. Becklin, I. Gatley and M. W. Werner, Astrophys. J. **258** 134 (1982).
5. D. F. Lester, M. W. Werner, J. W. V. Storey, D. M. Watson and C. H. Townes, Astrophys. J. (Letters) **248**, L109 (1981).
6. R. Genzel, D. M. Watson, C. H. Townes, D. F. Lester, H. L. Dinerstein, M. W. Werner and J. W. V. Storey, The Galactic Center, eds. G. Riegler and R. Blanford (American Institute of Physics, N. Y., 1982), p. 72.
7. H. S. Liszt, W. B. Burton and J. M. Van der Hulst, Astron. Astrophys. **142**, 237 (1985).
8. R. Gusten, R. Genzel, M. C. H. Wright, D. T. Jaffe, J. Stutzki and A. I. Harris, Preprint (1986).
9. Aa. Sandqvist, Astron. Astrophys. Suppl. **9**, 391 (1973).
10. R. Genzel, D. M. Watson, C. H. Townes, H. L. Dinerstein, D. Hollenbach, D. F. Lester, M. W. Werner and J. W. V. Storey, Astrophys. J. **276**, 551 (1984).
11. Aa. Sandqvist, A. Wootten and R.B. Loren, Astron. Astrophys. **152**, L25 (1985).
12. I. Gatley, T. J. Jones, A.R. Hyland, R. Wade, T.R. Geballe and K. Krisciunas, Mon. Not. R. Astr. Soc. **222**, 299 (1986).
13. E. Serabyn, R. Gusten, C. M. Walmsley, J. E. Wink and R. Zylka, Preprint (1987).

LOW-LUMINOSITY SEYFERT NUCLEI IN NEARBY GALAXIES

Alexei V. Filippenko
Department of Astronomy, University of California, Berkeley, CA 94720

Wallace L. W. Sargent
Palomar Observatory, Calif. Institute of Technology, Pasadena, CA 91125

ABSTRACT

Broad Hα emission lines have been detected in the nuclei of many bright galaxies such as M81 and M87. These lines are similar to, but much weaker than, those seen in type 1 Seyfert nuclei and QSOs. *If* massive black holes are responsible for the broad lines and the immense luminosities of classical AGNs, continuity arguments suggest that they also produce the features observed at lower levels in the relatively "normal" galaxies studied here, especially since the intensity ratios of the *narrow* emission lines are like those expected from gas photoionized by dilute nonstellar (e.g., power-law) radiation. The possible presence of massive black holes in nearby galaxies is consistent with the conclusion that the nucleus of our own Milky Way may harbor such an object.

INTRODUCTION

During the past three years, we have been engaged in a major observational effort to discover and study nearby, low-luminosity ("dwarf") active galactic nuclei (AGNs)[1,2]. We are obtaining spectra of every galaxy with $\delta \geq 0°$ and $B_T \leq 12.5$ mag, and carefully inspecting these data for evidence of Seyfert-like activity. We assume that a Seyfert 1 nucleus is characterized by the presence of Hα emission, no matter how faint, which is substantially broader than strong forbidden lines such as [O III] λ5007 and [N II] λ6583; typically this means that the full-width near zero intensity (FWZI) of Hα is $\gtrsim 4000$ km s^{-1}. The relative intensities of the narrow emission lines, often the only visible lines, constitute secondary criteria that can be used to distinguish between weak Seyfert 2 nuclei, Heckman's "low-ionization nuclear emission-line regions" (LINERs)[3], H II regions, and other types of objects.

Roughly 500 galaxies satisfy our selection criteria. Spectra of ~ 400 have already been obtained, but not yet fully analyzed. The Double Spectrograph[4] at the Cassegrain focus of the Hale 5.08-m reflector at Palomar Observatory is being used to procure the two-dimensional (long slit) spectra, which are recorded on charge-coupled devices.

M81: THE NEAREST KNOWN SEYFERT 1 GALAXY

Figure 1 shows an optical spectrum of the nucleus of M81, obtained on 25 February 1986 UT. The data have moderate resolution (~ 1.6 – 4 Å) and exceptionally high signal-to-noise ratios ($\gtrsim$ 100:1). The broad component of Hα emission first noticed by Peimbert and Torres-Peimbert[5] is easily visible; M81 therefore harbors a type 1 Seyfert nucleus, the nearest ($d \approx 3.3$ Mpc) discovered to date. Strong forbidden lines are also present, but many of the features were severely contaminated by the underlying starlight in the original spectrum. An absorption-line template galaxy, NGC 4339, was used to eliminate the stars, revealing the pure emission-line spectrum; see ref. 6.

In addition to the broad component of Hα emission, which has FWZI ≈ 6900 km s^{-1}, the data also convincingly show, for the first time, the corresponding component of Hβ (FWZI ≈ 5900 km s^{-1}). After using the [S II] λ6731 profile as a template to accurately remove the [N II] λλ6548, 6583 lines and the narrow component of Hα (ref. 6), we measure a full-width at half maximum (FWHM) of ∼ 2200 km s^{-1} for the broad Hα emission. The luminosity of this line is ∼ 1.2 × 10^{39} ergs s^{-1}, about 0.05 that of Hα in the nucleus of NGC 4051, the faintest classical type 1 Seyfert[7]. Since the 2 − 10 keV luminosity was ∼ 3 × 10^{40} ergs s^{-1} in early 1985 (ref. 8), we derive $L_X/L_{H\alpha}$ ≈ 25 if $L_{H\alpha}$ was roughly similar in February 1986 and early 1985. This is somewhat less than the nominal value (∼ 40) in bright AGNs[9]; hence, M81 is not "overluminous" at X-ray energies, suggesting that X-ray binaries within ∼ 20 pc of the nucleus contribute little to the X-ray flux. The intensity ratio of broad Hα to broad Hβ is ∼ 6, rather than the Case B recombination ratio of ∼ 2.9; if reddening is the sole reason for the discrepancy, an extinction A_V ≈ 2 mag is calculated. Under the assumption that the observed width of Hα is produced by clouds in Keplerian orbits, the mass enclosed by the broad-line region is calculated to be ∼ 5 × 10^5 $M_\odot$. The method of Wandel and Yahil[10] used to derive this number, however, may not be valid; we quote a formal mass only for comparison with those of other AGNs discussed in ref. 10.

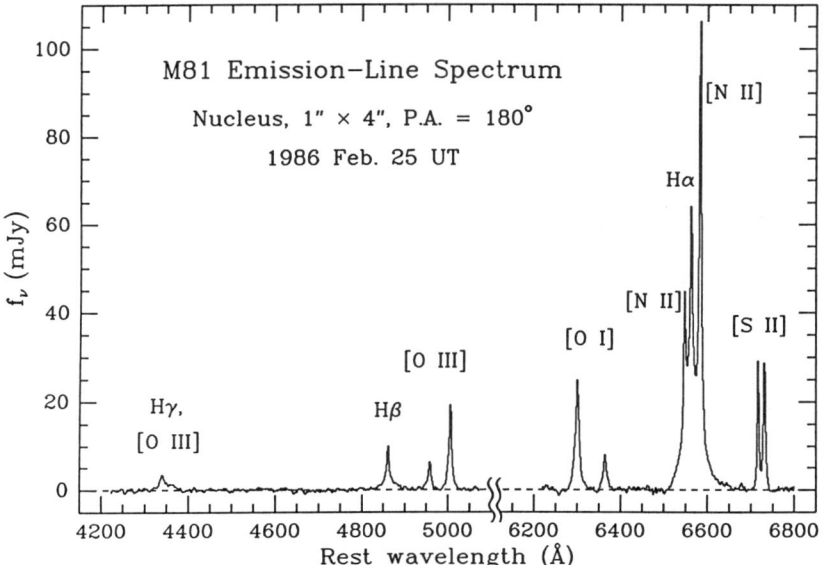

Figure 1: Flux-calibrated, unsmoothed spectrum of M81. A suitably scaled spectrum of NGC 4339, an absorption-line galaxy, has been subtracted from that of M81. Easily visible are the broad component of Hα and Hβ emission, as well as the markedly different widths and profiles of the unblended [O I], [O III], and [S II] lines. M81 exhibits classical LINER characteristics, with the strengths of [O I], [S II], and [N II] comparable to that of [O III] λ5007.

As demonstrated for certain other galaxies[1,2,11-13], the forbidden lines in M81 exhibit a strong correlation between profile width and critical density for collisional de-excitation; [O I] λ6300, for example, is far broader than either of the [S II] $\lambda\lambda$6716, 6731 lines, and its wings are much more extended (FWZI $\approx$ 2100 km s^{-1}). This implies that the narrow-line region of M81 is composed of clouds having a wide range of electron densities, from well below 10^3 cm^{-3} to well above 10^7 cm^{-3}. The densest clouds are optically thick, have the highest bulk velocities, and probably live closest to the nucleus[11]. Additional support for this interpretation is found from a detailed analysis of the individual [S II] lines themselves[6]: [S II] λ6716 is noticeably narrower than [S II] λ6731, whose critical density is a factor of $\sim$ 2.6 larger. The relative intensity of the two lines actually reaches the high-density limit ($n_e \gtrsim 10^5$ cm^{-3}) in the extreme wings, which are produced by high-velocity gas.

Comparison of spectra obtained over several days, one month, and two years with the same telescope, spectrograph, and entrance aperture reveals no substantial changes in the shape or strength of the broad Hα line[6]. Given its very low luminosity, this is rather puzzling, especially since M81 sometimes exhibits great X-ray variability (factor of $\sim$ 2 in $\sim$ 10 min; ref. 8). It is possible that the intrinsic luminosity of the continuum remains constant, and that *rapid* X-ray fluctuations are caused by broad-line clouds moving through the line of sight. This sort of mechanism, however, might still be expected to produce long-term variations in the line profiles.

OTHER LOW-LUMINOSITY SEYFERT NUCLEI

Spectra of many additional galaxies having broad Hα emission were shown at the Symposium. These will be discussed, but not illustrated, here.

An object whose classification as a Seyfert 1 is beyond doubt is NGC 4639 (spectrum in ref. 2), which exhibits very strong, broad Hα emission (FWZI $\approx$ 8600 km s^{-1}, FWHM $\approx$ 3700 km s^{-1}). NGC 4639 is located in the Virgo cluster, and may have been overlooked in previous surveys for AGNs because the broad component of Hβ is very weak. In this way its spectrum resembles that of NGC 4235 (ref. 1), which is also a member of the Virgo cluster. The broad Hα line in NGC 4639 is more prominent than that of any other galaxy in our survey, with the exception of previously recognized Seyfert 1 galaxies.

Several other objects exhibit easily visible wings on either side of the (Hα + [N II]) blend. An example is NGC 3998, whose broad Hα emission has FWZI $\gtrsim$ 8000 km s^{-1} (refs. 1, 2). As in M81, [O I] λ6300 is broader than the [S II] lines and has wings that are considerably stronger than those in a Gaussian profile. Subtle, but nevertheless perceptible, broad components of Hα have been found in about 10% of all galaxies, excluding dwarfs. Two particularly important objects are M87 (refs. 1, 2), whose nucleus has long been suspected to contain a supermassive black hole[14,15], and NGC 1052 (ref. 1), the prototypical LINER[16]. The luminosity of the broad Hα emission is generally $\sim 10^{38} - 10^{40}$ ergs s^{-1}. A correlation between the widths and critical densities of forbidden lines is present in many of the galaxies having broad Hα emission, and in over 40% of Heckman's original list[3] of LINERs.

Although the spectral characteristics of LINERS were at one time attributed to shocks[3,16], during the past few years considerable evidence has accumulated in support of photoionization by a nonstellar (e.g., power-law) continuum of the type found in QSOs and classical AGNs. In LINERs, however, the ratio of ionizing photons to nucleons (the "ionization parameter") is thought to be small[17,18]. A few notable discrepancies exist between observations and the predictions of photoionization models, but these largely disappear when the high range of densities found among the narrow-line clouds is included in the calculations[1,2,11-13]. Thus, it is likely that photoionization accounts for the observed relative intensities of the narrow lines in a large fraction of bright, nearby galaxies. The source of the ionizing radiation may be quite similar to that in QSOs. This leads us to believe that the broad Hα emission in these objects is also indicative of low-level QSO activity, rather than of some other phenomenon.

If classical AGNs are ultimately powered by supermassive black holes ($10^6 \lesssim M/M_\odot \lesssim 10^{10}$), then a significant fraction of local galaxies may now harbor these objects. Of course, it may also be that the black holes in most nearby galactic nuclei have much lower masses, and that the activity we have found was *always* very weak. Another possibility is that black holes are not responsible for the emission lines in a majority of galaxies; we note that there is currently *no* unequivocal evidence for massive black holes in *any* galaxy. Nevertheless, the possible presence of massive black holes in many nearby galaxies is consistent with the hypothesis, expressed numerous times during this Symposium, that our own Milky Way might contain such a beast ($M \lesssim 5 \times 10^6 \, M_\odot$). The Hubble Space Telescope could someday provide us with the direct evidence we need for at least a few external galaxies.

We thank the Time Allocation Committee of Palomar Observatory for large quantities of observing time. This research is supported by CalSpace grant CS-21-86 to A. V. F., and by NSF grant AST 84-16704 to W. L. W. S.

REFERENCES

1. A. V. Filippenko and W. L. W. Sargent, *Ap. J. Suppl.*, **57**, 503 (1985).
2. A. V. Filippenko and W. L. W. Sargent, in *Structure and Evolution of Active Galactic Nuclei*, eds. G. Giuricin *et al.* (Reidel: Dordrecht, 1986), p. 21.
3. T. M. Heckman, *Astr. Ap.*, **87**, 152 (1980).
4. J. B. Oke and J. E. Gunn, *Pub. A.S.P.*, **94**, 586 (1982).
5. M. Peimbert and S. Torres-Peimbert, *Ap. J.*, **245**, 845 (1981).
6. A. V. Filippenko and W. L. W. Sargent, *Ap. J.*, submitted (1987).
7. K. S. Anderson, *Ap. J.*, **162**, 743 (1970).
8. P. Barr and R. F. Mushotzky, *Nature*, **320**, 421 (1986).
9. M. Elvis, A. Soltan, and W. C. Keel, *Ap. J.*, **283**, 479 (1984).
10. A. Wandel and A. Yahil, *Ap. J. (Letters)*, **295**, L1 (1985).
11. A. V. Filippenko and J. P. Halpern, *Ap. J.*, **285**, 458 (1984).
12. A. V. Filippenko, *Ap. J.*, **289**, 475 (1985).
13. A. V. Filippenko, in *IAU Symposium 119, Quasars*, eds. G. Swarup and V. K. Kapahi (Reidel: Dordrecht, 1986), p. 289.
14. P. J. Young, J. A. Westphal, J. Kristian, C. P. Wilson, and F. P. Landauer, *Ap. J.*, **221**, 721 (1978).
15. W. L. W. Sargent, P. J. Young, A. Boksenberg, K. Shortridge, C. R. Lynds, and F. D. A. Hartwick, *Ap. J.*, **221**, 731 (1978).
16. J. A. Baldwin, M. M. Phillips, and R. Terlevich, *Pub. A.S.P.*, **93**, 5 (1981).
17. G. J. Ferland and H. Netzer, *Ap. J.*, **264**, 105 (1983).
18. J. P. Halpern and J. E. Steiner, *Ap. J. (Letters)*, **269**, L37 (1983).

A MAGNETIC LOOP MODEL FOR ACTIVITY IN THE GALACTIC CENTRE

Jean Heyvaerts
Observatoire de Paris, F-92190 Meudon

Ralph E. Pudritz
Dept of Physics, McMaster Univ., Hamilton, Ontario L8S 4M1

Colin A. Norman
Dept. of Physics and Astronomy, The Johns Hopkins Univ, Baltimore, MD 21218

ABSTRACT

We propose that radio structures on 0.1-100 pc scales in the galactic centre radio lobe (GCL) are manifestations of magnetic activity in a central source object. The observations indicate that truly one dimensional structures occur and we hold that this is strong evidence for magnetic loops. They are generated on subpc scales and expand out to 10-100 pc. Their interaction with the 2 pc molecular torus as well as the GCL on 30-50 pc scales reproduces the observed irregular ionized and molecular emission on 2 pc scales, as well as the bridge, radio arc, and thread-like filaments on the larger ones. The Sgr A radio lobe itself (on 10 pc scales) is comprised of a system of lower energy magnetized loops which have been trapped in the molecular torus. The main thrust of this theory is that all of these exotic structures may be understood within the framework of one rather simple model.

INTRODUCTION

The galactic centre harbours remarkably filamentary and ordered structures on scales within 100 pc of IRS16. The GCL has been modelled as a stubby, radio emitting shell with 'walls' 16 pc thick[1,2] and the 20 pc 'radio arc' portion near the galactic plane has been resolved into a series of parallel, slightly curved filaments[3]. Cutting nearly perpendicular to these structures are the more highly curved 'bridge' filaments which seem to originate in the outer portions of Sgr A. This complex itself contains filamentary-like protrusions[4]. This trend is seen again at 1-2 pc scales where an east-west 'bar' of ionized gas emission appears to cross and interact with gas comprising the torus of ionized emission[5,6] which may be associated with the molecular torus[7]. An important clue to the nature of these structures is the apparent braiding of the radio continuum emission on 2 pc scales[8]. A detailed model for this emission in terms of streams of tidally disrupted gas[9] has no possibility of producing braided gas motions. This feature is readily explained in our model however because the magnetic loops have a fairly strong, associated toroidal field component. The steep velocity gradient observed at the intersection of the bar with the torus also has no explanation in such tidal disruption models. The expansion of a magnetic loop into a magnetised torus however must undergo a reconnections event with produces high speed gas motions which explains the observation. Similarly, the collision of a loop, expanding out to 30 pc, with the magnetised GCL leads to localized particle injection. The resulting highly polarized emission then peaks where the current loop/GCL interaction is occurring, in agreement with the observations. The magnetic flux which must be associated with the observed filaments places interesting constraints on the central emitting body. The minimum mass body which can trap the huge magnetic flux which one may deduce from rotation measure observations[10] requires a mass of at least $10^5 M_\odot$ inside 10^{-2} pc. Plausible sources of activity include IRS16, the interaction of IRS16 and Sgr A*, or perhaps a disk around a central compact object. The ultimate source of mechanical energy for loop generation is a strongly sheared flow. The physics of loop eruption and subsequent propagation in an external medium has a well studied analogue on the sun[11]. It is known that large coiled magnetic loops erupt off the sun once per month and that their signatures are observed up to a distance of at least $10^2 R_\odot$.

THE MODEL

Three basic assumptions in this picture are: 1. The GCL is a cylindrical discontinuity separating an interior, hot, possibly X-ray emitting gas from an exterior, strongly magnetized, neutral (possibly molecular) component. 2. Filamentary 'braided' structures inside the GCL are portions of magnetic loops. This includes the radio bridge (30 pc), Sgr A (10 pc) and the 'East-West Bar' (2 pc). 3. Central loop generator produces loops powerful enough to expand out to 30-100 pc.

Loop Generation: A shear gradient on the central body twists a magnetic tube, creating a braided field. The loop detaches from the source after one relative rotation of the footpoints.

Loop Evolution: Loop expansion is driven by radial Lorentz forces produced by a poloidal loop current. This current is generated by the twisting of field lines while the loop is attached. We use a modification of Anzer's model[12] for solar coronal transients in writing down the model loop evolution equation. The mass carrying loops are retarded by (a) galactic gravitational potential and (b) aerodynamic drag by hot gas. The potential from 1-100 pc scales was modified from recent rotation curve observations[13]. The two main parameters of the model are (i) the ratio of loop magnetic to gravitational potential energy, γ_M and (ii) the ratio of mass pushed aside by loop expansion to loop mass, γ_D, which measures drag effects.

Loop Expansion Speed as a Function of Size: Numerical solutions to the loop evolution equation for the case of a massive compact object, $M_{hole} \sim M_{clus} \sim 10^6 M_\odot$ and the case of a cluster dominated central potential, $M_{hole} \simeq 10^{-2} M_{clus} = 10^4 M_\odot$ were found[14]. The drag coefficient is of order $\gamma_D = .002$ (this value deriving from the observed hot component pressure) but higher values for propagation through a higher pressure medium than observed ($\gamma_D = .01$.) were also used. The results show that peak expansion speeds are reached on slightly sub pc scales (source at 0.1 pc), that strong magnetic eruptions required to reach 30 pc, and that much higher loop velocities are achieved in models not dominated by a central hole (note, radial expansion force $\propto r^{-2}$). The best match to recombination line velocity data is for high mass: $\gamma_M = 0.8, \gamma_D = 0.002$, which produces a peak speed of 318 km s^{-1} and 184 km s^{-1} at the GCL (30 pc); and for low mass: $\gamma_M = 0.4, \gamma_D = 0.002$, with a peak speed of 313 km s^{-1} and 110 km s^{-1} at the GCL. Less energetic eruptions than these produce a population of loops that may fill 2-15 pc scales characteristic of Sgr A. If expanded 50pc scale loops are to be produced, the field strength at the source must be

$$B_s \leq 0.76 \, M_{s,6} \left(\frac{r_s}{.07 \text{pc}}\right)^{-2} \text{G}$$

LOOP-TORUS INTERACTION

A super-Alfvenic shock will form where a loop, expanding from sub pc scales, encounters the molecular torus. Reconnection of the loop with torus field produces the very broad lines and steep velocity gradient at points where the loop cuts the ionized inner edge of the torus. The highest velocity gas moves at the loop Alfven speed

$$v_A = 316 \left(\frac{B_l}{10^{-2} \text{ G}}\right) \left(\frac{n_l}{4 \times 10^3}\right)^{0.5} \text{ km s}^{-1}$$

High energy particles and waves propagate away from the shock at the torus Alfven speed

$$v_{A,tor} = 60 \left(\frac{B_{tor}}{10^{-2} \text{ G}}\right) \left(\frac{n_{tor}}{10^5 \text{ cm}^{-3}}\right)^{-1/2} \text{ km s}^{-1}$$

Thus, the disturbance created by the loop/torus shock is carried 'downwind' by the rotation of the torus (v_{rot}=100 km s^{-1}) at 160 km s^{-1}. The mechanical energy dumped into the torus is 1.5×10^{36} erg. Mass is deposited on torus, which exceeds $5 \times 10^3 M_\odot$ i.e., the torus can be built up by loop born mass deposition in about 10^7 yr.

The line widths of v=55 km s^{-1} observed in the molecular torus[7] are clearly sub-Alfvenic in the torus. By balancing the mechanical energy input into the torus by colliding loops, with the damping rate, we find that the line width is at least 43 km s^{-1} with the expansion speeds of 300 km s^{-1} inferred for our loops on this scale[14]. Loops that are trapped by the torus gradually cool and their gas drains back down to their footpoints in the torus. Because of their strong magnetic fields, these magnetized gas clumps will strongly resist being sheared and may retain their integrity for some time, as the observations demand.

SGR A AS A CORONA

Magnetic loops of lower energy have their footpoints embedded 2-5 pc torus. The turbulent dissipation of D.C. currents in these in the loops gives a heating rate of order[15]

$$H = 1.1 \times 10^{37} \left(\frac{B_{tor}}{10^{-2} \text{ G}}\right)^2 \left(\frac{R_{tor}}{2 \text{ pc}}\right)^2 \left(\frac{v_{shear}}{55 \text{ km s}^{-1}}\right) \text{ erg s}^{-1}$$

This can ultimately result in driving a wind of total energy 10^{52} erg if heating continues for more than 10^7 years. This may be sufficient to create the GCL itself, as non-radiative winds can be more efficient than supernovae in creating the cavity.

LOOP-GCL INTERACTION

The loop-GCL interaction produces a copious flux of shear Alfven waves on the strongly magnetized exterior to the GCL. The colliding loop has small rotation while the GCL rotates at 110 km s^{-1}. The two points on the GCL north and south of the galactic plane where a loop is merging produce strong heating and particle acceleration, as well as accounting for strong polarized emission at higher galactic latitudes than the points where bridge filaments cross the Radio Arc. For a GCL density of 10 cm^{-3}, the GCL field strength is .25 milliGauss. Alfven waves propagate at an angle of 32° from vertical, producing line of sight field of .16 milligauss, which accounts for Rotation Measure of 'polarized lobe' emission. The wave flux is sufficient to heat the gas to $T_{GCL} = 9000°K$. The Alfven velocity in the magnetised GCL is about 180 km s^{-1} so that the loop/GCL collision will only give rise to a low Mach number shock at best. The loop may actually penetrate the GCL under these conditions and this is the explanation for the curved filaments to the east of the radio arc. This penetration process gives rise to an electromotive force which accelerates electrons away from the two points on the GCL at which bridge filaments are merging. If 1% of the population is accelerated, one can produce the luminosity of the polarized radio lobe by a population of $\gamma = 10^3$ electrons. The current system set up on the GCL is aligned with the flux of Alfven waves flowing up the GCL.

Radio Arc Filaments: The GCL is concave outwards according to this model and has hot, possibly X-ray emitting gas on the interior and cool neutral gas exterior. The magnetic field is nearly perpendicular to the galactic plane. This configuration of the GCL is Rayleigh-Taylor unstable and long flute like modes will develope in a time

$$\tau = 1.7 \times 10^4 \left(\frac{R}{30 \text{ pc}}\right)^{0.5} \left(\frac{a_l}{1 \text{ pc}}\right)^{0.5} \left(\frac{T_x}{10^7 \text{ K}}\right)^{-0.5} \text{ years}$$

where T_x is the hot gas temperature, R is the curvature of the GCL and a_l is the perturbation scale, taken to be characteristic of the loop. Because the loop is the source of GCL perturbations, this instability will grow quickest on all portions of the GCL which has been affected by the loop collision. Hence, the radio arc filaments may be pictured as a rippled curtain whose fluted 'peaks' appear as the GCL straight parallel filaments. The loop/GCL interaction produces a plume-like system of waves and currents on the GCL which explains the the observed emission. We predict that the GCL emission to the west of Sgr A is also produced by a loop collision, which may have been a slightly earlier eruption. It would be important to search for the analogue of 'bridge' filaments to verify this hypothesis.

CONCLUSIONS

The idea that one dimensional loops are being shed from some central active source, hidden inside 0.05 pc of the centre shows that the structures in the galactic centre do not arise from tidal disruption and infall of tidally disrupted clouds, but are rather consequences of outflow and expansion. We show elsewhere[14] that a detailed accounting can be made of the ionized and molecular tori on 2 - 5 pc scales as well as the bridge, radio arc, and polarized emissions on scales characteristic of the GCL. The generator mass must be at least $10^5 M_\odot$ but no constraints can as yet be placed on whether a massive compact object is involved. The model shows however that less extreme conditions are required for loop propagation if a possible black hole is of low mass.

We are particularly grateful to Peter Quinn for several stimulating discussions. JH was supported by the INSU and the Observatoire de Paris, and thanks the Johns Hopkins University for hospitality and financial assistance. REP and CAN were supported by an IRAS Data Processing Grant No 957687.

REP acknowledges additional support of a Canadian NSERC grant 5-37687.

REFERENCES

1. Y. Sofue and T. Handa, Publ. Astron. Soc. Japan **36**, 539 (1984).
2. Y. Sofue, Publ. Astron. Soc. Japan **37**, 697 (1985).
3. F. Yusef-Zadeh, M. Morris, and D. Chance, Nature **310**, 557 (1984).
4. F. Yusef-Zadeh and M. Morris, preprint, Columbia Univ. (1986).
5. R.D. Ekers, J.H. van Gorkom, U.J. Schwarz, and Goss, W.M., Astron. Astrophys. **122**, 143 (1983).
6. K-.Y. Lo and M.J. Claussen, Nature **306**, 647 (1983).
7. R. Gusten, R. Genzel, M.C.H. Wright, D.T. Jaffe, and J. Stutzki, preprint, Radio Astronomy Lab, Berkeley (1986).
8. K-Y. Lo, preprint, Caltech Owen's Valley, (1986).
9. P.J. Quinn and G.J. Sussman, Ap. J. **288**, 377 (1985).
10. M. Tsuboi, M. Inoue, J.. Handa, H. Tabara, T. Kato, Y. Sofue, and N. Kaifu, preprint, Nobeyama Radio Observatory, (1987).
11. W.J. Wagner, Ann. Rev. Astron. Astrophys.**22** , 267 (1984).
12. U. Anzer, Solar Physics **57** , 111 (1978).
13. J.B. Lugten, R. Genzel, M.K. Crawford, and C.H. Townes, Ap. J. **306**, 691 (1986).
14. J. Heyvaerts, R.E. Pudritz, and C.A. Norman, in preparation (1986).
15. J. Heyvaerts and E.R. Priest, Astron. Astrophys. **117**, 220 (1983).

FURTHER "LOSS OF WEIGHT" BY A BLACK HOLE AT THE GALACTIC CENTER

L. M. Ozernoy
Harvard-Smithsonian Center for Astrophysics
Cambridge, MA 02138

ABSTRACT

I summarize here some recent and new arguments which constrain the mass of a black hole at the Galactic center by a value not exceeding a few hundred solar masses.

INTRODUCTION

The usual approach to obtain some basic parameters of a black hole (BH) presumably located at the Galactic center, such as the BH mass M_h and the accretion rate $\dot M$, is based on two fundamental assumptions: 1) When obtaining $M_h \sim 5 \cdot 10^6 M_\odot$ by using observational data on the kinematics of the ionized gas clouds in the central parsec of the Galaxy, one usually assumes these clouds to be rotating around the center including the innermost region[1], which seems to be hardly probable. 2) In order to avoid the accretion luminosity to be greater that the observed one, one supposes $\dot M$ to be very small, which requires its having an exceptionally small α-parameter[2, and refs. therein]. A possibility of self-consistent realization of a stable ion-supported torus with very small α and $\dot M$ is not yet proven.

Instead of attempts to derive M_h and $\dot M$ under special additional assumptions, I consider here possible ways of obtaining some constraints to such a fundamental parameter as M_h.

UPPER LIMIT TO M_h IMPOSED BY MASS OUTFLOW FROM IRS 16

Observations[3] of broad HeI and HI lines from the central near-infrared source IRS 16 indicated strongly an outflow of matter from it. Recent data[4] on the molecular hydrogen emission observed at the inner edge of the molecular ring confirmed convincingly an earlier explanation[5] of this emission by shocks due to mass loss from the center. If this wind is caused by outflow of matter from an accreting black hole (which seems to be quite natural in the framework of the BH concept applied to the Galactic center) one can use it to constrain the BH mass[6,7]. Indeed, in order to produce a wind, any BH model requires $L/L_{Edd} \gtrsim 1$, L_{Edd} being the usual Eddington luminosity (one neglects for a moment by $e^\pm$ pair production). This inequality, when using the observed total (mostly infrared) luminosity of the central source, yields immediately

$$M_h \lesssim 300 M_\odot. \qquad (1)$$

Creation of pairs makes each proton effectively coupled to $(1 + 2n_+/n_i)$ electrons (if pairs are prevented from escaping before they annihilate) so that the Eddington limit it reduced[8] to $L^*_{Edd} = L_E(1 + 2n_+/n_i)^{-1}$ where n_+/n_i is a function of many different parameters, including the ratio ϵ/μ, ϵ being the efficiency of accretion, and μ the inward drift time scale in units of the free-fall time. One can readily show that in the (realistic) limit of very large electron compactness parameter, L^*_E differs only slightly from L_E whenever $\epsilon/\mu \lesssim 0.02$, which seems to be rather reasonable for spherical or quasi-spherical accretion. Therefore, accounting for the pair-reducing Eddington limit can increase a little the above upper limit to M_h.

Moreover, using the size of the region where the $e^{\pm}$ pairs are created one can obtain an even more stringent (although model-dependent) upper limit to M_h.

CONSTRAINT TO M_h FROM $e^{\pm}$ PAIR CREATION

A promising mechanism of explaining $e^{\pm}$ pair creation in the Galactic center is an electromagnetic cascade within radiation-dominated plasma[9, and refs. therein]. Assuming that the hard continuum radiation up to $E \sim 150\ KeV$ from the Galactic center is due to electromagnetic cascade one can obtain the upper limit to the source's size at which the development of the cascade is still effective ($\tau_{\gamma\gamma} \gg 1$). In order to provide both the necessary generation rate of positrons, $Q_+ \simeq 2.10^{43} s^{-1}$, and the luminosity in soft gamma rays, $L_\gamma(E_\gamma \leq 2MeV) \simeq 2.10^{38} erg/s$, the source radius, according to Monte-Carlo calculations, should be $R \simeq 3.10^7 (1+\tau_{es})$ cm, τ_{es} being the optical depth in Thompson scattering, and the factor $(1+\tau_{es})$ accounts for the "wandering" of X-ray quanta in the source until escape[10]. As the cascade develops closely to the gravitatonal radius R_g of the accreting black hole, from the requirement $R \geq R_g$ where $R \simeq (3-10)10^7$ cm at $\tau_{es} = 0-2$ one gets

$$M_h \lesssim (10-30) M_\odot \qquad (2)$$

— a remarkably stringent upper limit to the BH mass.

AN ESTIMATE OF M_h USING DISPLACEMENT OF SGR A* FROM IRS 16

It was suspected a few years ago[11,12] that the compact radio source Sgr A* commonly identified with the hypothetical BH is located about 0.5 arcsec off the center of ionized gas clouds (the "mini-spiral") and IRS 16. Recent data[13] based on a more accurate technique confirmed non-coincidence of Sgr A* with either component of IRS 16. It has been proposed[14] that the possible displacement of Sgr A* be used for estimating the BH mass.

A massive BH placed in the center of a dense stellar system such as a galactic nucleus should move in response to gravitational pertubations from the stars in the nucleus core[15], mostly from interactions with unbound stars[16]. These interactions tend to bring the BH's kinetic energy into equilibrium with the kinetic energy of the passing stars, and the time needed to do this is typically very short as compared with the nucleus age. The mean value of the projected distance of the BH from the center of the core of the radius r_c is[15,16] $<d> = (\pi m_*/6M_h)^{1/2} r_c$ so that one can estimate the BH mass to be given by

$$M_h \simeq 100 \left(\frac{\theta}{0.''5}\right)^{-2} \left(\frac{r_c}{0.5\,\text{pc}}\right)^2 \frac{m_*}{0.5 M_\odot} M_\odot, \qquad (3)$$

θ being the observable angular distance between the BH and the Galactic center. Available uncertainties of quantities entering eq. (3) imply, in accordance with ref. 13, that $M_h \lesssim 100 M_\odot$ could be considered as a conservative upper limit to the BH mass at the Galactic center. Further observations using the Space Telescope can improve considerably this limit.

CONCLUSIONS

The upper limits (1), (2) and (3) to the black hole mass at the Galactic center confirm and even strengthen the limits obtained earlier by other methods[17,18]. A detailed theoretical analysis of the gradual growth of a seed black hole embedded into a dense star nucleus shows that under a rather wide range of initial conditions (core density n_* and stellar velocity dispersion v_*) the seed BH of $M_h \sim (10-100)M_\odot$ grows during the Hubble time very slowly[19,20]. Formation of a much more massive BH needs higher initial n_* and v_*, which could be realized in AGN's and QSO's, objects with quite different evolutionary histories.

REFERENCES

1. M.K. Crawford, R. Genzel, A.I. Harris, D.T. Jaffe, J.H. Lacy, J.B. Lutgen, E. Serabyn, and C.H. Townes, Nature, 315, 467(1985).
2. M. Rees, in The Galactic Center, G.R. Riegler and R.D. Blandford, Eds. (Amer. Inst. of Phys. Conf. Proc. 83, AIP, New York) p. 166(1982).
3. T.R. Geballe, K. Krisciunas, T.J. Lee, I. Gatley, R. Wade, W.D. Duncan, R. Garden, and E.E. Becklin, Astrophys. J., 284, 118(1984).
4. I. Gatley, T.J. Jones, A.R. Hyland, R. Wade, T.R. Geballe, K. Krisciunas, M.N.R.A.S., 222, 299(1986).
5. I. Gatley, T.J. Jones, A.R. Hyland, D.H. Beattie, and T.J. Lee, M.N.R.A.S., 210, 565(1984).
6. L.M. Ozernoy, Astron. Tsirk., no. 1342, 1(1984).
7. L.M. Ozernoy, in Twelfth Texas Symposium on Relativistic Astrophysics, M. Livio and G. Shaviv., Eds. Ann. New York Acad. of Sci., 470, 385(1986).
8. A.P. Lightman and A.A. Zdziarski, Preprint (1986).
9. F.A. Aharonian and L.M. Ozernoy, Astrophys. and Space Sci., (1987).
10. G.B. Rybicki and A.P. Lightman, Radiative Processes in Astrophysics, Wiley, New York, NY, (1979).
11. J.M.V. Storey and D.A. Allen, M.N.R.A.S., 204, 1153(1983).
12. R.L. Brown and H.L. Liszt, Ann. Rev. Astron. Astrophys. 22, 223(1984).
13. D.A. Allen and R.H. Sanders, Nature, 319, 191 (1986).
14. L.M. Ozernoy, "On the Nature of The Compact Source at the Galactic Center," paper presented at the XVIII General Assembly of IAU (1982), read by R. Ramaty.
15. J.N. Bahcall and R.A. Wolf, Ap. J. 209, 214(1976).
16. D.N.C. Lin and S. Tremaine, Ap. J. 242, 789(1980).
17. L.M. Ozernoy, in The Large-Scale Characteristics of the Galaxy, (Proc. IAU Symp. No. 84, ed. W.B. Burton), Dordrecht-Holland, p. 395(1979).
18. V.G. Gurzadyan and L.M. Ozernoy, Astron. Astrophys., 86, 315(1980).
19. L.M. Ozernoy, in Supermassive Black Holes, M. Kafatoc, Ed., Cambridge University Press (1987).
20. L.M. Ozernoy and V.I. Dokuchaev, in preparation.

POSITRON ANNIHILATION IN THE GALACTIC CENTER: "CHESHIRE CAT" COMPTON SCATTERING AND "EXCESS CONTINUUM"

M. Lars Bildsten
Cornell University, Department of Physics, Ithaca NY 14853

W. H. Zurek
Los Alamos National Laboratory, Theoretical Astrophysics
Los Alamos, NM 87545

ABSTRACT

Two separate observations of the γ-ray spectrum originating from the Galactic Center were made by HEAO-3 in the fall of 1979 and in the spring of 1980. The 2γ 511 KeV annihilation line flux decreased by a factor of three over the corresponding six month period, whereas the excess γ-ray continuum below the 511 KeV line, often interpreted as 3γ decay of orthopositronium, barely changed. This apparent discrepancy in the temporal behavior makes it difficult to associate the bulk of the excess continuum with the 3γ decay of positronium. We will show that Compton scattering of the line and high energy radiation provides a natural explanation for the surprisingly small change seen in the excess continuum.

INTRODUCTION

The HEAO-3 observations of the Galactic Center in the fall of 1979 and the spring of 1980 established this region as a strong source of annihilation radiation.[1] The flux in the narrow 511 KeV e^+e^- annihilation line reached $\sim 10^{43}$ annihilations per second. The energy of the line was almost exactly $m_e c^2$, limiting the gravitational redshift to less than 10^{-3}. The line intensity decreased by a factor of ~ 3 over a six month period. Therefore, the annihilation region must be smaller than 10^{18} cm.[2] A hydrogen column density of $\rho_H \sim 0.2 \text{g/cm}^2$ is needed to stop a relativistic (1 MeV) positron. Using this fact and the upper limit on the size of the region we obtain a lower limit on the number density $n_H > 10^5$ cm^{-3}.

In both a neutral and ionized medium, most of the positrons annihilate via the intermediate stage of a Positronium atom(Ps).[3,4] Positronium decays from its ground state, which can be either a spin triplet(parallel spins) or a spin singlet(antiparallel spins). The spin singlet (para-Ps) decays to two 511 KeV photons, whereas the spin triplet (ortho-Ps) decays to a 3γ continuum extending to 511 KeV. The linewidth of the 511 KeV line is determined by the average velocity of the Ps atom prior to annihilation. The observed linewidth limits the cloud temperature to $T_{cloud} < 10^4$K. Observations of the inner parsecs of the Galactic Center – extensively discussed in this volume – support the existence of HII regions with such temperatures and with number densities around 10^5 cm^{-3}.

The γ-ray continuum was simultaneously measured during the HEAO-3 observations. It has been modeled by a combination of a power-law, an ortho-Ps (3γ) decay continuum, and a Comptonized thermal emission spectrum to match the observed high energy component.[5] The fall '79 data show a high luminosity in photons of energy greater than 511 KeV. We shall refer to this portion of the spectrum as the "high-energy continuum". Photons in the energy range of 200-511 KeV, which cannot be accounted for by the extrapolation of low and high energy portions of the spectrum, are then regarded as an "excess continuum". This excess continuum was often regarded as evidence for 3γ decays of ortho-Ps. For annihilations proceeding exclusively through Ps, the ratio of 3γ continuum to 2γ line intensity is 9/2. Yet, the value obtained from the fall '79 data was much less than this. One of us has demonstrated that annihilation events occur predominantly on dust grains in an interstellar HII region containing the usual abundance of dust.[6] The lifetime of the ortho-Ps atom which forms on the grain is longer than the collision time with electrons in the dust grain. Therefore, the ortho-Ps atom undergoes collisional charge exchange prior to annihilation. This exchange converts the ortho-Ps to a spin singlet para-Ps, thereby suppressing the total number of 3γ decays and explaining the low observed flux associated with the excess continuum.

Any changes in the flux of the 511 KeV line should be accompanied by similar changes in the flux associated with ortho-Ps decay. But, the spring '80 observations showed that while the line flux dropped by a factor of three, the flux associated with the three photon decay decreased only slightly. Therefore the initial interpretation of the excess continuum in terms of ortho-Ps decay radiation is in doubt. Moreover, the high energy($E > 511$ KeV) component of the spectrum was no longer observed in the spring '80 data.

ROLE OF COMPTON SCATTERING: THE "CHESHIRE CAT" CONTINUUM

The parameters derived earlier for the annihilation region give a Compton scattering optical depth for 511 KeV photons of $\tau_{511} > 0.05$. Therefore, at least 5 percent of the line photons are Compton scattered in these HII regions on their way to Earth. The photon always loses energy in these scatterings, since the temperature of these regions is low compared to the photons' incident energy ($kT_{cloud} \ll 511$ KeV). Hence, the scattered line radiation falls in the energy range 170-511 KeV and contributes to the excess continuum. The spectrum of Compton scattered line radiation mimics that of 3γ decay, as originally pointed out by Forrest.[7] Scattered line photons arrive at Earth at a later time than unscattered ones originating from the same point. Therefore if the positron source suddenly turned off, the observer would initially only see a reduction in the line intensity. The flux in the excess continuum below the line would not show appreciable changes until the crossing time ($t_{cross} \sim R/c$) of the photon through the annihilation region has passed. This time delay, the "Cheshire Cat" effect, explains the lack of immediate changes in the excess continuum. If one assumes that the high energy component of the spectrum ($E > 511$ KeV) originates from the region of positron production, then it must also undergo Compton scattering. In view of the similar temporal variations of the line and the high energy component, this seems reasonable. The luminosity in the high energy continuum is approximately 10 times greater than the luminosity in the line. Therefore, even for low optical depths, Compton scattering of the line plus the high energy component can naturally account for the intensity as well as for the delay in temporal variations of the excess continuum. If the optical depth is indeed so low, Compton scattering of the line alone would affect the observations only very little. Nevertheless, when one includes the high-energy component enough photons are placed in the 200-511 KeV range to account for the intensity of the excess continuum. Moreover, it is also conceivable – although we shall disregard this possibility in the remainder of this note – that the optical depth is large enough so that Compton scattered line photons could account for most of the excess continuum.

MODEL

Realistic models of the Galactic Center annihilation source are not available at present. Therefore, we will work with a simple scenario, hoping that the salient features are independent of geometry. We place the photon source at the center of a spherical cloud of constant number density n_H. The problem is then one of deriving the scattered spectrum seen by the observer as a function of time. Given the past observations of the annihilation line, it is reasonable to use a source model that was constant for a long time and then died off with a six month decay time.[8] The observed spectrum of the excess continuum is then a function of the time since turnoff and the optical depth. The analytic calculation was done with a 0 or 1 scattering approximation and was compared to a Monte Carlo code to verify the results. We have made no attempt to fit the data with this model; this issue will be addressed elsewhere.[9] Here we merely want to establish the importance of Compton scattering in the Galactic Center annihilation region.

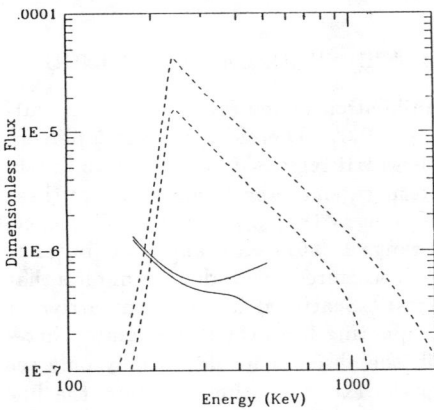

Figure 1: "Cheshire Cat" continua:
1) Solid line – Compton scattered 2γ annihilation photons. Note the absence of higher energy radiation in the lower curve which corresponds to the later time.
2) Broken line – Compton scattered high energy continuum. Prominent backscatter peak appears at 220 KeV.

This calculation was done for both the line and the high energy continuum. The spectra of scattered radiation seen by the observer for these two components are shown in Figure 1. Both of the top lines correspond to a steady state(fall '79), whereas the lower lines correspond to a time six months after turnoff(spring '80). The graphs show the relative constancy of the excess continuum. To see how the "Cheshire Cat" continuum affects the γ-ray observations, we added the broken line spectra of Figure 1 to the other two continuum components used by Riegler et. al.[5] The spectra in Figure 2 therefore represent what would be observed if the H.E. continuum was being Compton scattered since this dominates over the line alone being scattered. The power law(P) is determined by the low energy data. The high energy component(T) was taken to be the Comptonized thermal emission derived by Sunyaev and Trümper, which is similar, but not identical to that used by Riegler.[10,11] As in Riegler et. al's analysis, the H.E. continuum was taken to have 10 times the luminosity of the line. Note the backscatter peak at 220 KeV in both the fall and the spring. These plots should be regarded only as a qualitative illustration of the phenomena.

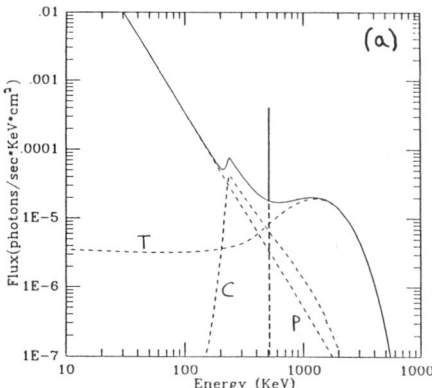 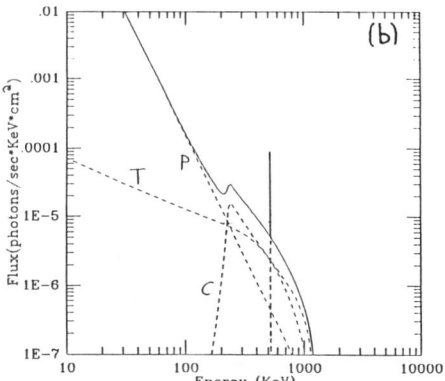

Figure 2: The evolution of the Galactic Center γ-ray spectrum predicted by the "Cheshire Cat" continuum model. The form of the fall '79(Figure 2a) and spring '80(Figure 2b) spectra was obtained by assuming that the H.E. continuum originated from a cloud of optical depth ~ 0.2 and radius ~ 2 light years.

CONCLUSIONS

The role of Compton scattering in the Galactic Center cannot be ignored. It provides a very natural explanation for the two separate HEAO-3 observations. This would also imply significant annihilation on dust since the 3γ decays must be substantially suppressed to obtain the observed time variation of the spectrum. The strongest prediction that we can make is that when the source turns back on, the excess continuum should be much weaker than in the steady state and must be largely attributed to 3γ decay, since Compton scattered photons would be delayed. Further observations are clearly needed, in addition to a reanalysis of the data including the effect of Compton scattering. In particular, it may be worthwhile to reexamine the data to search for the backscatter peak near 220 KeV.

REFERENCES

1. G. R. Riegler, J. C. Ling, W. A. Mahoney, W. A. Wheaton, J. B. Willett, A. S. Jacobson and T. A. Prince, Ap. J. Lett., 248, L13 (1981).
2. R. Ramaty and R. E. Lingenfelter, Phil. Trans. R. Soc. Lond., **A301**, 671 (1981).
3. B. L. Brown and M. Leventhal, Phys. Rev. Lett., **57**, 1651 (1986).
4. R. W. Bussard, R. Ramaty and R. J. Drachman, Ap. J., **228**, 928 (1979).
5. G. R. Riegler, J. C. Ling, W. A. Mahoney, W. A. Wheaton and A. S. Jacobson, Ap. J. Lett., **294**, L13 (1985).
6. W. H. Zurek, Ap. J., **289**, 603 (1985).
7. D. J. Forrest, *The Galactic Center*, ed. G. R. Riegler and R. D. Blandford (AIP, N.Y., 1982) p. 160.
8. C. J. MacCallum and M. Leventhal, *Positron-Electron Pairs in Astrophysics*, ed. M. L. Burns, A. K. Harding and R. Ramaty (AIP, N.Y., 1983) p. 211.
9. M. L. Bildsten and W. H. Zurek, Los Alamos preprint (1987).
10. R. A. Sunyaev and J. Trümper, Nature, **279**, 506 (1979).
11. R. A. Sunyaev and L. G. Titarchuk, Astr. Ap., **86**, 121 (1980).

MOLECULAR GAS ASSOCIATED WITH THE GALACTIC CENTER ARC

E. Serabyn and R. Güsten

Max-Planck-Institut für Radioastronomie
Auf dem Hügel 69, D-5300 Bonn 1, F.R.G.

The galactic center arc[1,2] has recently been shown to be composed of several thin, elongated filaments[3]. A number of almost straight, parallel filaments run perpendicular to the galactic plane at 1 - 0.17°, and several 'arched' filaments lie in the 'bridge' region, between 1 - 0.17° and the galactic center. Radio recombination line observations[2,4] reveal that the arched filaments are thermal in nature, while the straight filaments are non-thermal. The nature of these peculiar gas structures is not clear, but molecular line observations[5,6] show the existence of molecular gas with similar negative velocities in the bridge region. In order to determine if and how this molecular gas is associated with the thermal arched filaments, a high spatial resolution (25") molecular line map of the bridge region has been made in the CS J=2-1 rotational transition at 97.981 GHz, using the IRAM 30m telescope on Pico Veleta, Spain.

The resulting CS J=2-1 data show a clear association of molecular material with the arched filaments, both kinematically and spatially. The velocities of both gas components tend to increase from ~ -50 to ~ 0 km s^{-1} in moving northward along the filaments. Furthermore, a spatial superposition of the molecular and ionized gas components (fig. 1) reveals that the arched filaments lie primarily along the edges of molecular clouds. Given this intimate relationship between the gas components, it is clear that the structure of the arched filaments results from the distribution of molecular material. Models relying solely on ionized gas dynamics are thus ruled out. The molecular data suggest a model consisting of a tidally disrupted molecular cloud which is falling toward the galactic center[7].

REFERENCES

1. Pauls, T., Downes, D., Mezger, P.G. and Churchwell, E., 1976, A.A. 46, 407
2. Pauls, T. and Mezger, P.G., 1980, A.A. 85, 26
3. Yusef-Zadeh, F., Morris, M. and Chance, D., 1984, Nature 310, 557
4. Yusef-Zadeh, F., priv. comm.
5. Bieging, J., Downes, D., Wilson, T.L., Martin, A.H.M. and Güsten, R., 1980, A.A. Suppl. 42, 163
6. Bally, J., Stark, A.A., Wilson, R.W. and Henkel, C., 1986, preprint
7. Serabyn, E. and Güsten, R., 1987, preprint
8. Morris, M., and Yusef-Zadeh, F., 1985, Astronom. J. 90, 2511.

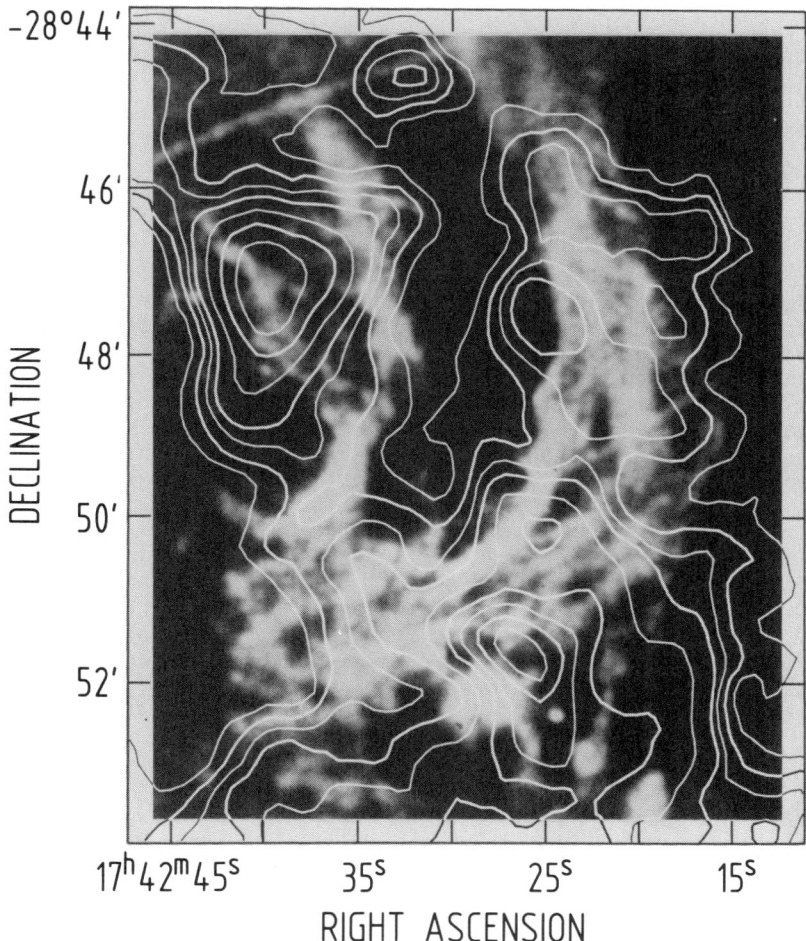

Figure 1. Contours of CS J=2-1 emission superposed on a 6 cm continuum radiograph of the arched filaments[8]. The CS contours give the integral of T_A^* in the range -55 to 5 km s^{-1}, with levels 10, 14, 18, 22, 30, 40, 50, 60, 70 K km s^{-1}.

SPATIAL AND KINEMATIC STRUCTURE OF THE THERMAL COMPONENTS OF THE GALACTIC CENTER ARC

F. Yusef-Zadeh
Columbia Univ., New York, NY 10027 and UCLA, Los Angeles, CA 90024

Mark Morris
UCLA, Los Angeles, CA 90024

J. H. van Gorkom
NRAO, Socorro, NM 87801

ABSTRACT

High-resolution radio continuum and radio recombination line observations of two bright segments of the filamentary Arc near the galactic center have been carried out using the VLA. On the basis of the polarization and recombination line characteristics of these regions, one can clearly identify and distinguish the thermal and non-thermal features of the Arc. These observations provide strong evidence in support of a physical interaction between these two components. In fact, the clarity of the interaction between thermal and non-thermal structures is unprecedented. In the regions of interaction, the magnetic field is dynamically important, and we infer that its strength exceeds 1 milligauss.

The dominant components of thermal emission are centered on G0.18-0.04 and G0.1+0.08, which are integral parts of the Arc, but appear to be physically coupled to two distinct molecular clouds -- the 50 and -30 km/s clouds. A complex and highly organized flow of ionized gas is seen in both sources. We consider several possibilities for the source of ionization in G0.18-0.04.

INTRODUCTION

Multiconfiguration radio continuum studies of the galactic center using the VLA reveal that the radio Arc located at l = 0.2° consists of a number of distinct radio features[1]. Three of the most important structures are: 1) a network of linear filaments which appear to maintain their spatial coherence over a scale of 40 pc, and which are oriented perpendicular to the galactic plane (see figure 4), 2) a system of arched filaments (G0.1+0.08) which meets the linear filaments at positive latitudes and is less uniform in its appearance than the system of linear filaments, and 3) a wispy, quasi-filamentary structure (G0.18-0.04) which appears to cross the linear filaments about where they intersect the galactic plane. On the basis of linear polarization and low-frequency measurements, it was previously shown that non-thermal emission arises from the linear filaments[2,3] and that the thermal gas, as evidenced by recombination line emission[4,5], lies in regions that roughly coincide with G0.1+0.08 and G0.18-0.04. Here, we confirm that these two HII regions are responsible for most of the thermal emission from the Arc; there is a detailed and close correspondence between the recombination line-emitting regions and the arched filaments and

G0.18-0.04. Also, each of these is associated in position and velocity with a molecular cloud.

OBSERVATIONS AND RESULTS

Radio recombination line emission (H110α) from the Arc was observed using the VLA in the hybrid C/D configuration. Fields at 6-cm were centered on both G0.1+0.08 and G0.18-0.04. A total bandwidth of 3.125 MHz was used, giving a velocity range of -80 to 110 km/s and a velocity resolution of 6 km/s. The phase and bandpass calibrators were 1748-253 and 2134+004, respectively.

Figures 1 and 2, which are based on multiconfiguration radio continuum observations using the VLA (see refs. 6 and 7 for observational details) depict the 6-cm images of G0.1+0.08 and G0.18-0.04.

In addition to showing the arched filaments, figure 1 shows in its northeastern corner segments of the most prominent non-thermal, linear filaments (a larger-scale perspective is given in figure 4). The thermal nature of the arched filaments is demonstrated in figure 3, which depicts the hydrogen recombination line emission (H110α). Almost all the emission in this region (1 ~ 0.1°) arises at negative velocities which are forbidden in the sense of galactic rotation. The velocities corresponding to the southern and western portions are generally more negative than their counterparts to the north and east, respectively. The negative-velocity gas in this region can also be noted in a map of ^{13}CO emission made by Bally et al.[8]. Figure 4 superimposes this map on the radio continuum data; it suggests that the molecular gas peaks at locations where the continuum emission is weak.

The high-resolution continuum image of G0.18-0.04 (fig. 2) shows a sickle-shaped feature crossing the network of nonthermal, linear filaments oriented perpendicular to the galactic plane. Almost all of the recombination line emission in this region arises from the sickle-shaped feature; none can be seen from the filaments. The H110α data indicate that, unlike the arched filaments, the entire emitting region has positive radial velocity, i.e., 30 - 50 km/s. A bright continuum spot at G0.15-0.05 ($\alpha = 17^h43^m05^s$, $\delta = -28°48'45"$) has the largest radial velocity in this region, > 110 km/s (lying just at the edge of the velocity coverage).

DISCUSSION

The relatively continuous velocity gradients along the spatially coherent thermal structures in the Arc imply a flow of ionized gas along them with a dynamical timescale of 2×10^5 years. The radial velocity gradient indicates that the flow is being accelerated or that the orientation of the filaments with respect to the line of sight varies along their length. If the velocity gradient were due to an enormous shear within the emitting region, we would expect the filaments to be a transient phenomena (lifetime ~ 2×10^5 years).

The southernmost of the linear, nonthermal filaments in figure 2 bends slightly as it crosses the sickle-shaped feature at $\alpha = 17^h42^m58.5^s$, $\delta = -28°48'08"$. This is the only unambiguous

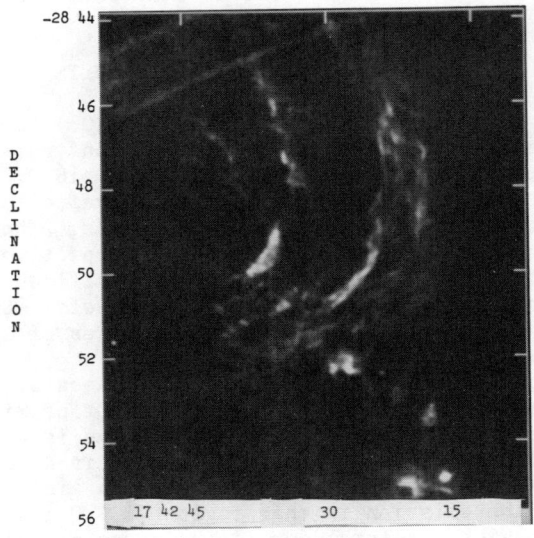

Figures 1 and 2: Radiographs of the 6-cm continuum emission from G0.1+0.08(above) and G0.18-0.04(below) with resolutions of 3.8"x3.1" and 2.6"x1.9", respectively.

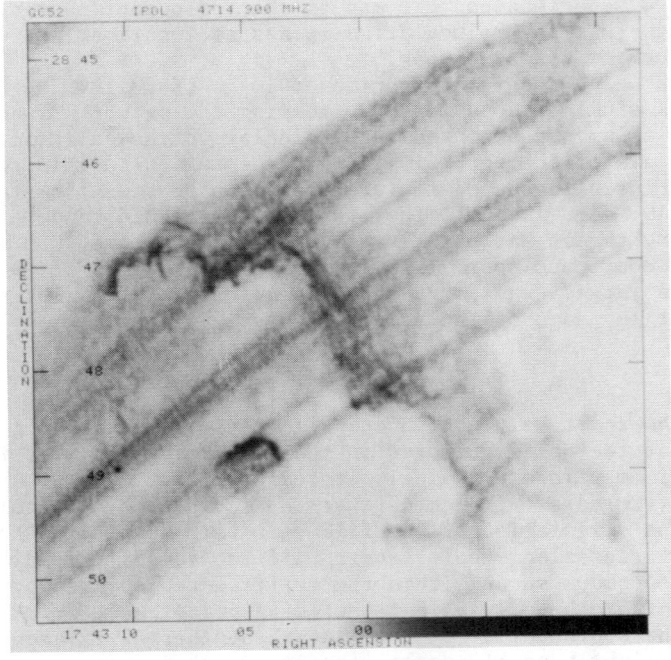

bending of any of the linear filaments in the system. It implies an interaction of the thermal and non-thermal features at this point. Further evidence in support of this hypothesis comes from several other straight filaments that abruptly change brightness as they pass through the sickle (see references 7 and 9 for more details). These changes in the brightness may imply that some of the energy contained in particles within the linear filaments is deposited in the sickle feature (see below). It is also possible that the magnetic field lines are illuminated as the bulk motion of the relativistic gas picks up and accelerates thermal electrons in G0.18-0.04.

The strength of the magnetic field in G0.18-0.04 is estimated to be ~ 0.2 milligauss if the magnetic field is assumed to be in pressure equilibrium with the thermal gas. The number density of electrons, n_e ~ 400 cm^{-3}, is estimated from the mean observed emission measure by assuming that the plasma has an electron temperature of 10^4 K. For G0.15-0.05, a higher value of 1 milligauss is obtained in the same manner. An even larger magnetic field is found if we note that the field lines in the linear filaments of the Arc (which are oriented along those filaments[10]) are only slightly bent, at most, upon interacting with the thermal gas. Estimating the relative velocity of the magnetic filaments and the thermal gas to be on the order of a few tens of km/s, we find that a magnetic field of at least a milligauss is needed to offset the ram pressure of the ambient thermal gas.

Because of the puzzling geometry and the complex kinematics of the Arc, neither the source(s) of ionization nor the origin of the non-thermal filaments have yet been identified. Here, we suggest a few possibilities that might account for the ionization and heating in G0.18-0.04:

1) A direct physical encounter of relativistic electrons in the linear filaments with ambient gas. The electrons could dump a portion of their energy and heat the ambient gas by Coulomb interaction. However, one would expect to observe large linewidths in the thermal gas, especially if the relativistic gas has a substantial bulk motion relative to the ambient gas. This is contrary to the observed 20 km/s recombination line widths.

2) Photoionization of thermal gas by UV photons emitted from the nonthermal filaments. This would require that the synchrotron radiation from the filaments has a flat spectrum up to ultraviolet frequencies, and thus that the relativistic γ factor for electrons is ~ 10^6. Recent spectral index maps of the Arc, which indicate that the Arc has a flat spectrum between 160 MHz and 45 GHz[3,11], imply γ ~ 3000, consistent with this possibility.

3) Yusef-Zadeh and Morris[9] have recently argued for a bulk streaming motion of relativistic electrons along the magnetic field lines. It is possible that the sickle-shaped structure seen in G0.18-0.04 is a shock front produced by the passage of relativistic gas through an inhomogeneous medium. This oblique shock front could also be produced by the supersonic motion of ambient gas across the network of magnetic filaments. The single filament which is

Figure 3: Radiographs of H110α recombination line emission. The LSR velocity of each of 12 adjacent channels is indicated in each box.

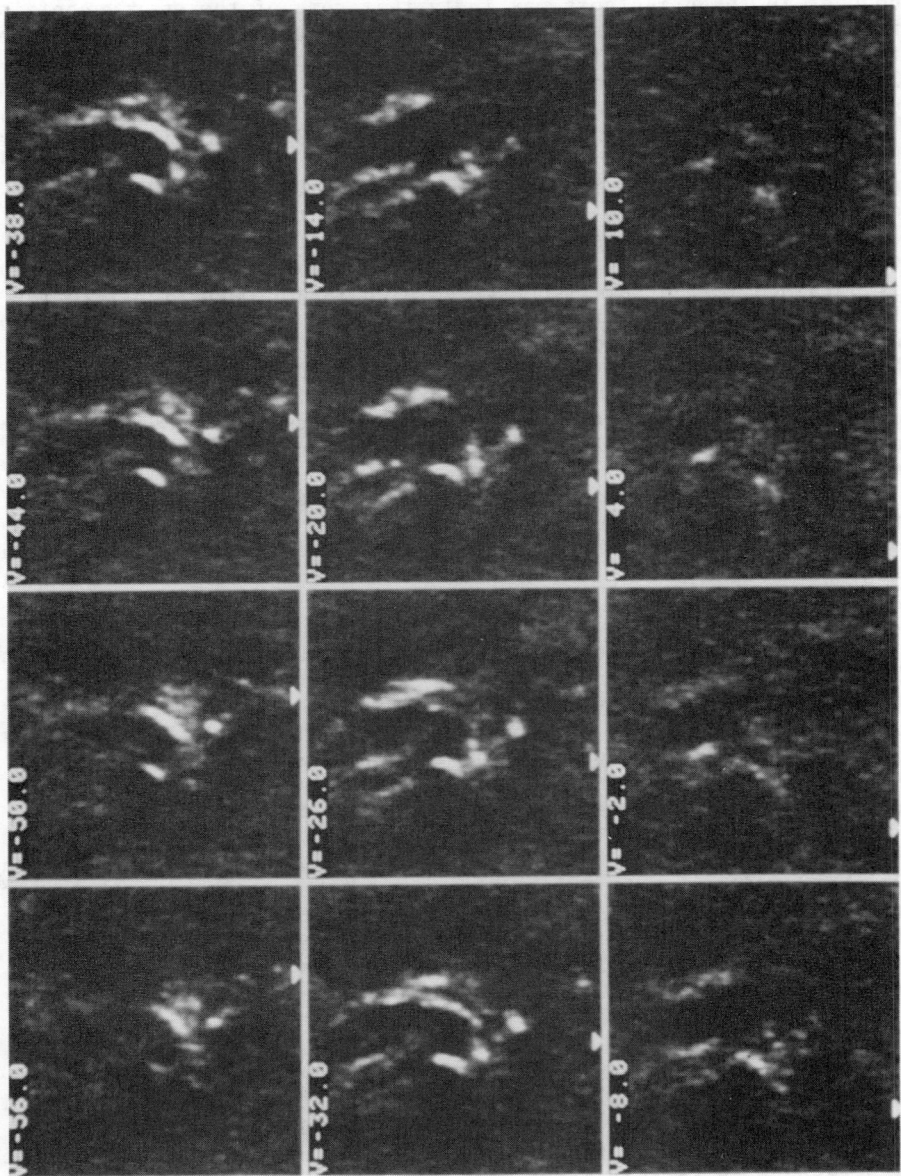

noticeably bent might then be explained as the "refraction" of magnetic field lines where they encounter an oblique switch-on shock in the sickle structure[12].

These heating mechanisms would not be appropriate for the arched filaments, which apparently have no non-thermal component. We discuss a magnetohydrodynamic model for the ionization of the arched filaments elsewhere.

REFERENCES

1. F. Yusef-Zadeh, M. Morris and D. Chance, Nature, 310, 557 (1984).
2. M. Inoue, T. Takahashi, H. Tabara, T. Kato and M. Tsuboi, Publ. Astr. Soc. Japan, 36, 633 (1984).
3. F. Yusef-Zadeh, M. Morris, B. Slee and G. Nelson, Ap. J., 310 (1986).
4. F. F. Gardner and J. B. Whiteoak, Proc. Astron. Soc. Australia, 3, 150 (1977).
5. T. A. Pauls and P. G. Mezger, Astron. Ap., 85, 26 (1980).
6. F. Yusef-Zadeh and M. Morris, submitted to Ap. J. (1986).
7. F. Yusef-Zadeh, Ph.D. Thesis, Columbia Univ. (1986).
8. J. Bally, A. A. Stark, R. Wilson and C. Henkel, submitted to Ap. J. (1986).
9. F. Yusef-Zadeh and M. Morris, in preparation (1987).
10. M. Tsuboi, M. Inoue, T. Handa, H. Tabara, T. Kato, Y. Sofue, and N. Kaifu, Astron. J., in press (1986).
11. Y. Sofue, M. Inoue, T. Handa, M. Tsuboi, H. Hirabayashi, M. Morimoto, and K. Akabane, P. A. S. Japan, in press (1986).
12. E. R. Priest, Solar Magnetohydrodynamics, Reidel (1984).

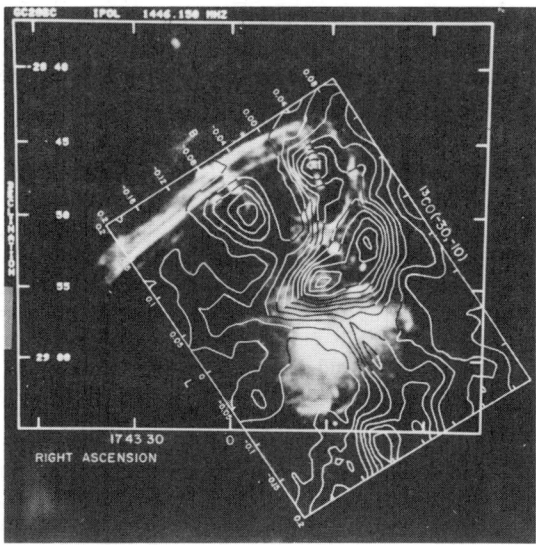

Figure 4: Contours of ^{13}CO J = 1→0 intensity integrated over the velocity range -10 to -30 km/s (ref. 8), superimposed on a 20-cm VLA radiograph.

Evidence For Activity At The Galactic Center Based On Low
Frequency Radio Continuum Observations

Namir E. Kassim and William C. Erickson
University of Maryland, College Park, Md. 20742

Theodore N. LaRosa
University of Alabama, Huntsville Al. 35899

Abstract

Aperture synthesis observations of the galactic center region at 123.0 and 110.6 MHz reveal striking asymmetric steep spectrum ($\alpha \leqslant -0.8$) radio lobes. The northern galactic lobe (NGL) appears to be a unique galactic source and cannot be easily classified as an SNR, extragalactic source, or a foreground object[1]. The southern "lobe" appears directly linked to the galactic center and has been identified by Yusef-Zadeh et al.[2] as a low energy jet emanating from Sgr-A. We present here new 80 MHz observations of an expanded region around the galactic center obtained one year after the initial set of observations. Excellent morphological agreement between these separate sets of observations confirm the persistence of the NGL as the dominant emission feature in the entire region at these frequencies. The similarities of this source with the jet feature suggest a common origin. Together with the compact nonthermal source located at the galactic center[3], the properties and geometry of these sources resemble the radio lobes observed in the nuclei of Seyfert galaxies[4].

Introduction

The Galactic center at metric wavelengths displays a complicated morphology which arises from the superposition of nonthermal emission processes and absorption due to ionized gas. We have made observations using the Clark Lake TPT telescope at 80 MHz[1] and at 110.6 and 123.0 MHz[5] with angular resolutions of 3 to 8 arc-minutes. Maps from these observations have been reproduced in Figs. 1, 2, and 3. The observations reveal steep-spectrum radio lobe features which appear to be associated with the galactic center.

In April 1985 we reobserved an expanded region around the galactic center at 80 MHz in order to confirm our earlier results and to search for additional unusual features which might be revealed on a larger map. These observations were conducted approximately one year after the initial 80 MHz observations.[1]

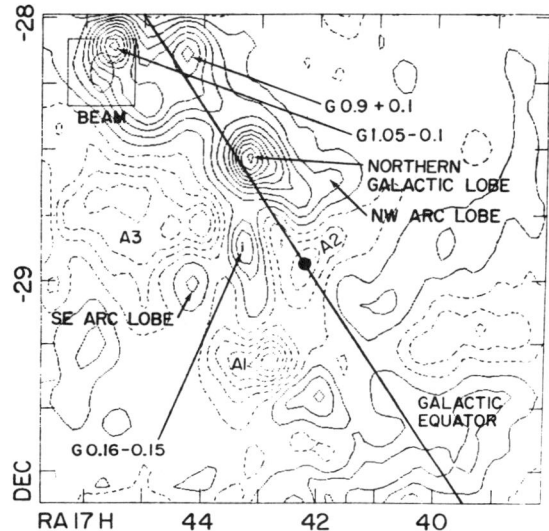

Fig. 1: Contour map at 80 MHz of the Galactic Center Region. Map center $17^h42^m30^s$, $-28°54'57"$; field $1.9°\times1.9°$; beamwidth $5.0'\times 8.6'$. Peak brightness temperature 43,750 K; first contour 2,300 K; contour interval 4,600 K. A1, A2, and A3 are absorption regions discussed previously[1]. A black dot indicates the position of the galactic center ($l \equiv 0$, $b \equiv 0$).

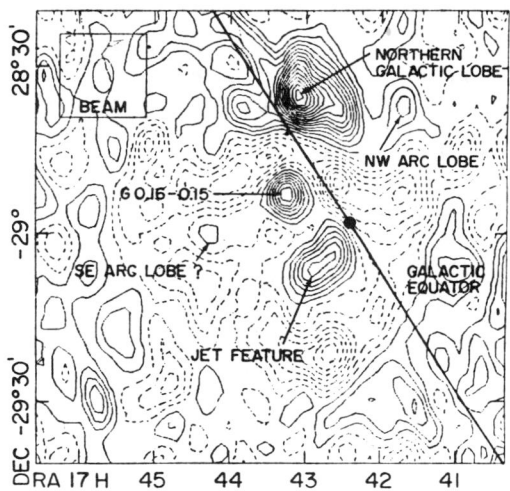

Fig. 2: Contour map at 110.6 MHz of the Galactic Center Region. Map center $17^h43^m30^s$, $-28°,59',57"$; field $1.4°\times1.4°$; beamwidth $3.6'\times 6.2'$. Peak brightness temperature 26,994 K; first contour 905 K; contour interval 1,795K. A black dot indicates the position of the galactic center ($l \equiv 0$, $b \equiv 0$).

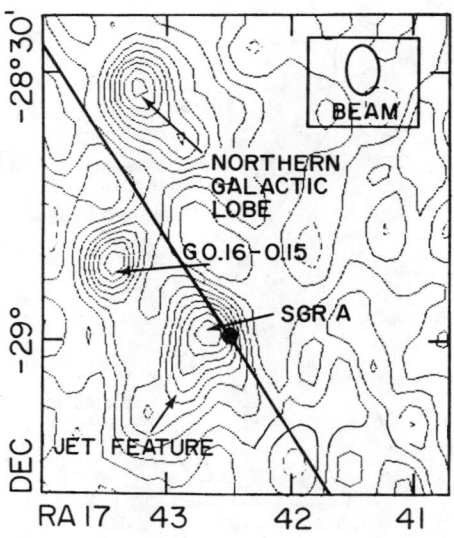

Fig. 3: Contour map of the Galactic Center Region at 123 MHz. Beamwidth 3.3´x5.8´; peak brightness temperature 54,600 K; first contour 2,753 K; contour interval 5,461 K. A black dot marks the position of the galactic center ($l \equiv 0$, $b \equiv 0$).

Observations and Reductions

The new 80 MHz data were processed as described earlier[1] and the final map is presented in Fig. 4. This map represents a composite map that was created by combining five separate maps made on five different nights. The field of view for each map at 80 MHz is approximately $2°x2°$ and the beamwidth in the direction of the galactic center is 5.0´x8.6´. The field centers for four of the maps were centered on the four corners of the original 80 MHz field of Fig. 1. A fifth map was centered as in Fig. 1. When combined this provided a composite field of view of approximately 16 square degrees.

Calibration of Fig. 4 based on observations of strong, small-diameter sources yielded a peak brightness temperature for the NGL which was 20% higher than the May-June 1984 observation[1]. We do not consider this variation significant.

Results and Discussion

The new observations confirm the previous ones. Three identified discrete emission features appear on Fig. 4, as expected. They are G0.9+0.1 and G1.05-0.1, probable SNRs, and G0.16-0.15 which is coincident with the nonthermal part of the continuum arc feature. The composite map also contains the HII region G1.1-0.1 which is seen in absorption. This source was also seen in absorption at 57.5 MHz as reported earlier[1]. The information available about the identified sources is summarized in Table 1.

Table 1: Identified Sources

G0.16-0.15: Coincident with the southeastern part of the continuum arc feature and evidence of nonthermal emission in this region of the arc.

G0.9+0.1: Most likely a galactic SNR, possibly a plerion.
α (15GHz-408MHz) $\sim$ -0.1;
α (408-80 MHz) $\sim$ -0.06

G1.05-0.1: Probable galactic SNR.
α (408-80 MHz) $\sim$ -0.4.

G1.1-0.1: An HII region seen in absorption against the galactic background.

The larger field of view provides a perspective which highlights the uniqueness of the NGL as well as the unusual tight clustering of the discrete sources along the galactic plane northward of the center. Previously[1,5] we reported three ways in which the relationship between the steep spectrum lobes and the compact nonthermal source located at the galactic center parallel behavior observed in active galaxies. They were: 1) The steep spectrum of the lobes compared to the spectrum of the compact source; 2) The ratio of the luminosity of the lobes to that of the compact source; 3) The spatial scale of the lobes and compact source. All three of these characteristics were confirmed for the NGL by the new observations and the results for both steep spectrum features and the compact nonthermal source are summarized quantitatively in Table 2.

Table 2
Observed properties of features located near the galactic center

	110.6 MHz flux density (Jy)	Luminosity (erg s^{-1})	α
Northern Galactic Lobe	65	2×10^{34}	$\leqslant -1.0$
Jet Feature	10	2×10^{33}	$\leqslant -0.7$
Central Compact Source	--	2×10^{34}	~ -0.25

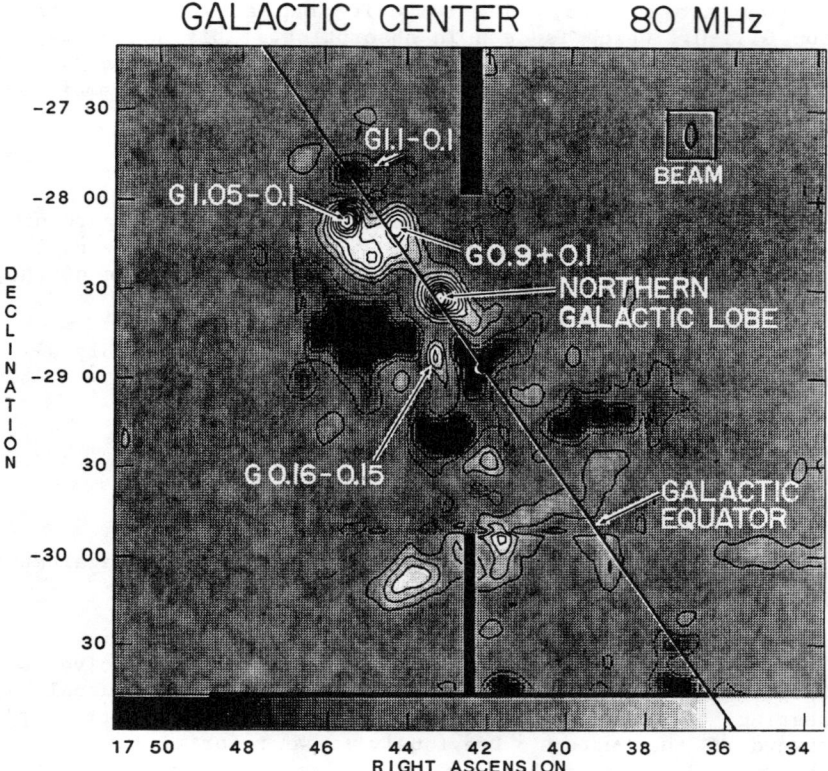

Fig. 4: Composite 80 MHz map of the Galactic Center Region. Field $3.8° \times 3.8°$; beamwidth $5.0' \times 8.6'$. A black dot indicates the position of the galactic center ($l \equiv 0$, $b \equiv 0$). Dark areas indicate regions of ionized gas seen in absorption against the distributed galactic background radiation.

Conclusions

The expanded field of the 80 MHz observations presented here shows that the previously observed NGL is the dominant source in this area and this provides further evidence that it is probably associated with the galactic center. The 110.6 and 123.0 MHz observations also leave no question of the existence of the jet feature discovered by Yusef-Zadeh et al. at 160 MHz. The chance of finding two extended, steep-spectrum radio sources within 30 arc-minutes of the galactic center is unlikely. In the absence of alternative hypotheses it seems that the NGL and southern jet feature are related and are a manifestation of activity at the galactic center similar to that observed in active galaxies.

References

1. LaRosa, T. N. and Kassim, N. E., Ap. J. (Letters) **299**, L13 (1985)

2. Yusef-Zadeh, F., Morris, M., Slee, O. B., and Nelson, G. J., Ap. J. (Letters) **300**, L47 (1986).

3. Lo, K. Y., Backer, D. C., Ekers, R. D., Kellerman, K. I., Reid, M., and Moran, J. M., Nature **315**, 124 (1985).

4. Wilson, A. S., in IAU Symposium 97, Extragalactic Radio Astronomy, ed. D. S. Heeschen and C. M. Wade (Dordrecht: Reidel, 1981), p. 179.

5. Kassim, N. E., LaRosa, T. N., and Erickson, W. C., Nature **322**, 522 (1986).

AUTHOR INDEX

A

Allen, D. A., 1
Arens, J. F., 142

B

Backer, D. C., 163
Barrett, A. H., 99
Becklin, E. E., 87, 162
Bildsten, M. L., 184

C

Churchwell, E., 110
Crawford, M. K., 123

D

Dinerstein, H., 162
Downes, D., 166

E

Ellis, H. B., 138
Erickson, W. C., 196
Evans, N. J., 114

F

Fazio, G. G., 146, 157
Filippenko, A. V., 172
Forrest, W. J., 153
Fukui, Y., 110

G

Gaalema, S., 142
Gardner, F. F., 95
Gatley, I., 8, 87, 106, 146, 162
Gautier, T. N., 157
Geballe, T. R., 39
Genzel, R., 103, 118, 123, 133, 166
Gezari, D. Y., 146
Güsten, R., 19, 103, 133, 188
Gwinn, C. R., 166

H

Hall, D. N. B., 83, 87
Harris, A. I., 103, 118

Harvey, P. M., 138
Hayashi, M., 106
Heyvaerts, J., 176
Ho, P. T. P., 99
Hoffmann, W. F., 146, 157

I

Inatani, J., 106

J

Jackson, J. M., 99
Jaffe, D. T., 103, 133
Jones, B., 162
Joy, M., 138

K

Kaifu, N., 106
Karlsson, R., 95
Kassim, N. E., 196
Kleinmann, S. G., 83
Koch, D. G., 157

L

Lacy, J. H., 142
Lamb, G., 146
LaRosa, T. N., 196
Lebofsky, M. J., 79, 91
Lester, D. F., 138, 142
Lingenfelter, R. E., 51
Lo, K. Y., 30
Low, F. J., 157
Lugten, J. B., 118, 123

M

McCreight, C., 146
McGinn, M. T., 87
Melnick, G. J., 157
Moran, J. M., 166
Morris, M., 127, 190

N

Norman, C. A., 176

O

Ozernoy, L. M., 181

P

Peck, M. C., 142
Pipher, J. L., 153
Pudritz, R. E., 176

R

Ramaty, R., 51
Rees, M. J., 71
Reid, M. J., 166
Rieke, G. H., 79, 91, 157
Rönnäng, B., 166

S

Sandqvist, A., 95, 168
Sargent, W. L. W., 62, 172
Schneps, M. H., 166
Scoville, N. Z., 83
Sellgren, K., 83, 87
Serabyn, E., 188
Shu, P., 146
Shure, M. A., 153
Sramek, R. A., 163

Stacey, G. J., 118
Stutzki, J., 103

T

Townes, C. H., 118, 123
Tresch-Fienberg, R., 146

V

van Gorkom, J. H., 190

W

Werner, M. W., 162
Whiteoak, J. B., 95
Woodward, C. E., 153
Wright, M. C. H., 103, 133

Y

Young, E. T., 157
Yusef-Zadeh, F., 127, 190

Z

Zurek, W. H., 184

AIP Conference Proceedings

		L.C. Number	ISBN
No. 1	Feedback and Dynamic Control of Plasmas – 1970	70-141596	0-88318-100-2
No. 2	Particles and Fields – 1971 (Rochester)	71-184662	0-88318-101-0
No. 3	Thermal Expansion – 1971 (Corning)	72-76970	0-88318-102-9
No. 4	Superconductivity in d- and f-Band Metals (Rochester, 1971)	74-18879	0-88318-103-7
No. 5	Magnetism and Magnetic Materials – 1971 (2 parts) (Chicago)	59-2468	0-88318-104-5
No. 6	Particle Physics (Irvine, 1971)	72-81239	0-88318-105-3
No. 7	Exploring the History of Nuclear Physics – 1972	72-81883	0-88318-106-1
No. 8	Experimental Meson Spectroscopy –1972	72-88226	0-88318-107-X
No. 9	Cyclotrons – 1972 (Vancouver)	72-92798	0-88318-108-8
No. 10	Magnetism and Magnetic Materials – 1972	72-623469	0-88318-109-6
No. 11	Transport Phenomena – 1973 (Brown University Conference)	73-80682	0-88318-110-X
No. 12	Experiments on High Energy Particle Collisions – 1973 (Vanderbilt Conference)	73-81705	0-88318-111-8
No. 13	π-π Scattering – 1973 (Tallahassee Conference)	73-81704	0-88318-112-6
No. 14	Particles and Fields – 1973 (APS/DPF Berkeley)	73-91923	0-88318-113-4
No. 15	High Energy Collisions – 1973 (Stony Brook)	73-92324	0-88318-114-2
No. 16	Causality and Physical Theories (Wayne State University, 1973)	73-93420	0-88318-115-0
No. 17	Thermal Expansion – 1973 (Lake of the Ozarks)	73-94415	0-88318-116-9
No. 18	Magnetism and Magnetic Materials – 1973 (2 parts) (Boston)	59-2468	0-88318-117-7
No. 19	Physics and the Energy Problem – 1974 (APS Chicago)	73-94416	0-88318-118-5
No. 20	Tetrahedrally Bonded Amorphous Semiconductors (Yorktown Heights, 1974)	74-80145	0-88318-119-3
No. 21	Experimental Meson Spectroscopy – 1974 (Boston)	74-82628	0-88318-120-7
No. 22	Neutrinos – 1974 (Philadelphia)	74-82413	0-88318-121-5
No. 23	Particles and Fields – 1974 (APS/DPF Williamsburg)	74-27575	0-88318-122-3
No. 24	Magnetism and Magnetic Materials – 1974 (20th Annual Conference, San Francisco)	75-2647	0-88318-123-1
No. 25	Efficient Use of Energy (The APS Studies on the Technical Aspects of the More Efficient Use of Energy)	75-18227	0-88318-124-X

No. 26	High-Energy Physics and Nuclear Structure – 1975 (Santa Fe and Los Alamos)	75-26411	0-88318-125-8
No. 27	Topics in Statistical Mechanics and Biophysics: A Memorial to Julius L. Jackson (Wayne State University, 1975)	75-36309	0-88318-126-6
No. 28	Physics and Our World: A Symposium in Honor of Victor F. Weisskopf (M.I.T., 1974)	76-7207	0-88318-127-4
No. 29	Magnetism and Magnetic Materials – 1975 (21st Annual Conference, Philadelphia)	76-10931	0-88318-128-2
No. 30	Particle Searches and Discoveries – 1976 (Vanderbilt Conference)	76-19949	0-88318-129-0
No. 31	Structure and Excitations of Amorphous Solids (Williamsburg, VA, 1976)	76-22279	0-88318-130-4
No. 32	Materials Technology – 1976 (APS New York Meeting)	76-27967	0-88318-131-2
No. 33	Meson-Nuclear Physics – 1976 (Carnegie-Mellon Conference)	76-26811	0-88318-132-0
No. 34	Magnetism and Magnetic Materials – 1976 (Joint MMM-Intermag Conference, Pittsburgh)	76-47106	0-88318-133-9
No. 35	High Energy Physics with Polarized Beams and Targets (Argonne, 1976)	76-50181	0-88318-134-7
No. 36	Momentum Wave Functions – 1976 (Indiana University)	77-82145	0-88318-135-5
No. 37	Weak Interaction Physics – 1977 (Indiana University)	77-83344	0-88318-136-3
No. 38	Workshop on New Directions in Mossbauer Spectroscopy (Argonne, 1977)	77-90635	0-88318-137-1
No. 39	Physics Careers, Employment and Education (Penn State, 1977)	77-94053	0-88318-138-X
No. 40	Electrical Transport and Optical Properties of Inhomogeneous Media (Ohio State University, 1977)	78-54319	0-88318-139-8
No. 41	Nucleon-Nucleon Interactions – 1977 (Vancouver)	78-54249	0-88318-140-1
No. 42	Higher Energy Polarized Proton Beams (Ann Arbor, 1977)	78-55682	0-88318-141-X
No. 43	Particles and Fields – 1977 (APS/DPF, Argonne)	78-55683	0-88318-142-8
No. 44	Future Trends in Superconductive Electronics (Charlottesville, 1978)	77-9240	0-88318-143-6
No. 45	New Results in High Energy Physics – 1978 (Vanderbilt Conference)	78-67196	0-88318-144-4
No. 46	Topics in Nonlinear Dynamics (La Jolla Institute)	78-57870	0-88318-145-2
No. 47	Clustering Aspects of Nuclear Structure and Nuclear Reactions (Winnepeg, 1978)	78-64942	0-88318-146-0
No. 48	Current Trends in the Theory of Fields (Tallahassee, 1978)	78-72948	0-88318-147-9

No. 49	Cosmic Rays and Particle Physics – 1978 (Bartol Conference)	79-50489	0-88318-148-7
No. 50	Laser-Solid Interactions and Laser Processing – 1978 (Boston)	79-51564	0-88318-149-5
No. 51	High Energy Physics with Polarized Beams and Polarized Targets (Argonne, 1978)	79-64565	0-88318-150-9
No. 52	Long-Distance Neutrino Detection – 1978 (C.L. Cowan Memorial Symposium)	79-52078	0-88318-151-7
No. 53	Modulated Structures – 1979 (Kailua Kona, Hawaii)	79-53846	0-88318-152-5
No. 54	Meson-Nuclear Physics – 1979 (Houston)	79-53978	0-88318-153-3
No. 55	Quantum Chromodynamics (La Jolla, 1978)	79-54969	0-88318-154-1
No. 56	Particle Acceleration Mechanisms in Astrophysics (La Jolla, 1979)	79-55844	0-88318-155-X
No. 57	Nonlinear Dynamics and the Beam-Beam Interaction (Brookhaven, 1979)	79-57341	0-88318-156-8
No. 58	Inhomogeneous Superconductors – 1979 (Berkeley Springs, W.V.)	79-57620	0-88318-157-6
No. 59	Particles and Fields – 1979 (APS/DPF Montreal)	80-66631	0-88318-158-4
No. 60	History of the ZGS (Argonne, 1979)	80-67694	0-88318-159-2
No. 61	Aspects of the Kinetics and Dynamics of Surface Reactions (La Jolla Institute, 1979)	80-68004	0-88318-160-6
No. 62	High Energy e^+e^- Interactions (Vanderbilt, 1980)	80-53377	0-88318-161-4
No. 63	Supernovae Spectra (La Jolla, 1980)	80-70019	0-88318-162-2
No. 64	Laboratory EXAFS Facilities – 1980 (Univ. of Washington)	80-70579	0-88318-163-0
No. 65	Optics in Four Dimensions – 1980 (ICO, Ensenada)	80-70771	0-88318-164-9
No. 66	Physics in the Automotive Industry – 1980 (APS/AAPT Topical Conference)	80-70987	0-88318-165-7
No. 67	Experimental Meson Spectroscopy – 1980 (Sixth International Conference, Brookhaven)	80-71123	0-88318-166-5
No. 68	High Energy Physics – 1980 (XX International Conference, Madison)	81-65032	0-88318-167-3
No. 69	Polarization Phenomena in Nuclear Physics – 1980 (Fifth International Symposium, Santa Fe)	81-65107	0-88318-168-1
No. 70	Chemistry and Physics of Coal Utilization – 1980 (APS, Morgantown)	81-65106	0-88318-169-X
No. 71	Group Theory and its Applications in Physics – 1980 (Latin American School of Physics, Mexico City)	81-66132	0-88318-170-3
No. 72	Weak Interactions as a Probe of Unification (Virginia Polytechnic Institute – 1980)	81-67184	0-88318-171-1
No. 73	Tetrahedrally Bonded Amorphous Semiconductors (Carefree, Arizona, 1981)	81-67419	0-88318-172-X

No. 74	Perturbative Quantum Chromodynamics (Tallahassee, 1981)	81-70372	0-88318-173-8
No. 75	Low Energy X-Ray Diagnostics – 1981 (Monterey)	81-69841	0-88318-174-6
No. 76	Nonlinear Properties of Internal Waves (La Jolla Institute, 1981)	81-71062	0-88318-175-4
No. 77	Gamma Ray Transients and Related Astrophysical Phenomena (La Jolla Institute, 1981)	81-71543	0-88318-176-2
No. 78	Shock Waves in Condensed Matter – 1981 (Menlo Park)	82-70014	0-88318-177-0
No. 79	Pion Production and Absorption in Nuclei – 1981 (Indiana University Cyclotron Facility)	82-70678	0-88318-178-9
No. 80	Polarized Proton Ion Sources (Ann Arbor, 1981)	82-71025	0-88318-179-7
No. 81	Particles and Fields –1981: Testing the Standard Model (APS/DPF, Santa Cruz)	82-71156	0-88318-180-0
No. 82	Interpretation of Climate and Photochemical Models, Ozone and Temperature Measurements (La Jolla Institute, 1981)	82-71345	0-88318-181-9
No. 83	The Galactic Center (Cal. Inst. of Tech., 1982)	82-71635	0-88318-182-7
No. 84	Physics in the Steel Industry (APS/AISI, Lehigh University, 1981)	82-72033	0-88318-183-5
No. 85	Proton-Antiproton Collider Physics –1981 (Madison, Wisconsin)	82-72141	0-88318-184-3
No. 86	Momentum Wave Functions – 1982 (Adelaide, Australia)	82-72375	0-88318-185-1
No. 87	Physics of High Energy Particle Accelerators (Fermilab Summer School, 1981)	82-72421	0-88318-186-X
No. 88	Mathematical Methods in Hydrodynamics and Integrability in Dynamical Systems (La Jolla Institute, 1981)	82-72462	0-88318-187-8
No. 89	Neutron Scattering – 1981 (Argonne National Laboratory)	82-73094	0-88318-188-6
No. 90	Laser Techniques for Extreme Ultraviolt Spectroscopy (Boulder, 1982)	82-73205	0-88318-189-4
No. 91	Laser Acceleration of Particles (Los Alamos, 1982)	82-73361	0-88318-190-8
No. 92	The State of Particle Accelerators and High Energy Physics (Fermilab, 1981)	82-73861	0-88318-191-6
No. 93	Novel Results in Particle Physics (Vanderbilt, 1982)	82-73954	0-88318-192-4
No. 94	X-Ray and Atomic Inner-Shell Physics – 1982 (International Conference, U. of Oregon)	82-74075	0-88318-193-2
No. 95	High Energy Spin Physics – 1982 (Brookhaven National Laboratory)	83-70154	0-88318-194-0
No. 96	Science Underground (Los Alamos, 1982)	83-70377	0-88318-195-9

No. 97	The Interaction Between Medium Energy Nucleons in Nuclei – 1982 (Indiana University)	83-70649	0-88318-196-7
No. 98	Particles and Fields – 1982 (APS/DPF University of Maryland)	83-70807	0-88318-197-5
No. 99	Neutrino Mass and Gauge Structure of Weak Interactions (Telemark, 1982)	83-71072	0-88318-198-3
No. 100	Excimer Lasers – 1983 (OSA, Lake Tahoe, Nevada)	83-71437	0-88318-199-1
No. 101	Positron-Electron Pairs in Astrophysics (Goddard Space Flight Center, 1983)	83-71926	0-88318-200-9
No. 102	Intense Medium Energy Sources of Strangeness (UC-Sant Cruz, 1983)	83-72261	0-88318-201-7
No. 103	Quantum Fluids and Solids – 1983 (Sanibel Island, Florida)	83-72440	0-88318-202-5
No. 104	Physics, Technology and the Nuclear Arms Race (APS Baltimore –1983)	83-72533	0-88318-203-3
No. 105	Physics of High Energy Particle Accelerators (SLAC Summer School, 1982)	83-72986	0-88318-304-8
No. 106	Predictability of Fluid Motions (La Jolla Institute, 1983)	83-73641	0-88318-305-6
No. 107	Physics and Chemistry of Porous Media (Schlumberger-Doll Research, 1983)	83-73640	0-88318-306-4
No. 108	The Time Projection Chamber (TRIUMF, Vancouver, 1983)	83-83445	0-88318-307-2
No. 109	Random Walks and Their Applications in the Physical and Biological Sciences (NBS/La Jolla Institute, 1982)	84-70208	0-88318-308-0
No. 110	Hadron Substructure in Nuclear Physics (Indiana University, 1983)	84-70165	0-88318-309-9
No. 111	Production and Neutralization of Negative Ions and Beams (3rd Int'l Symposium, Brookhaven, 1983)	84-70379	0-88318-310-2
No. 112	Particles and Fields – 1983 (APS/DPF, Blacksburg, VA)	84-70378	0-88318-311-0
No. 113	Experimental Meson Spectroscopy – 1983 (Seventh International Conference, Brookhaven)	84-70910	0-88318-312-9
No. 114	Low Energy Tests of Conservation Laws in Particle Physics (Blacksburg, VA, 1983)	84-71157	0-88318-313-7
No. 115	High Energy Transients in Astrophysics (Santa Cruz, CA, 1983)	84-71205	0-88318-314-5
No. 116	Problems in Unification and Supergravity (La Jolla Institute, 1983)	84-71246	0-88318-315-3
No. 117	Polarized Proton Ion Sources (TRIUMF, Vancouver, 1983)	84-71235	0-88318-316-1

No. 118	Free Electron Generation of Extreme Ultraviolet Coherent Radiation (Brookhaven/OSA, 1983)	84-71539	0-88318-317-X
No. 119	Laser Techniques in the Extreme Ultraviolet (OSA, Boulder, Colorado, 1984)	84-72128	0-88318-318-8
No. 120	Optical Effects in Amorphous Semiconductors (Snowbird, Utah, 1984)	84-72419	0-88318-319-6
No. 121	High Energy e^+e^- Interactions (Vanderbilt, 1984)	84-72632	0-88318-320-X
No. 122	The Physics of VLSI (Xerox, Palo Alto, 1984)	84-72729	0-88318-321-8
No. 123	Intersections Between Particle and Nuclear Physics (Steamboat Springs, 1984)	84-72790	0-88318-322-6
No. 124	Neutron-Nucleus Collisions – A Probe of Nuclear Structure (Burr Oak State Park - 1984)	84-73216	0-88318-323-4
No. 125	Capture Gamma-Ray Spectroscopy and Related Topics – 1984 (Internat. Symposium, Knoxville)	84-73303	0-88318-324-2
No. 126	Solar Neutrinos and Neutrino Astronomy (Homestake, 1984)	84-63143	0-88318-325-0
No. 127	Physics of High Energy Particle Accelerators (BNL/SUNY Summer School, 1983)	85-70057	0-88318-326-9
No. 128	Nuclear Physics with Stored, Cooled Beams (McCormick's Creek State Park, Indiana, 1984)	85-71167	0-88318-327-7
No. 129	Radiofrequency Plasma Heating (Sixth Topical Conference, Callaway Gardens, GA, 1985)	85-48027	0-88318-328-5
No. 130	Laser Acceleration of Particles (Malibu, California, 1985)	85-48028	0-88318-329-3
No. 131	Workshop on Polarized ^{3}He Beams and Targets (Princeton, New Jersey, 1984)	85-48026	0-88318-330-7
No. 132	Hadron Spectroscopy–1985 (International Conference, Univ. of Maryland)	85-72537	0-88318-331-5
No. 133	Hadronic Probes and Nuclear Interactions (Arizona State University, 1985)	85-72638	0-88318-332-3
No. 134	The State of High Energy Physics (BNL/SUNY Summer School, 1983)	85-73170	0-88318-333-1
No. 135	Energy Sources: Conservation and Renewables (APS, Washington, DC, 1985)	85-73019	0-88318-334-X
No. 136	Atomic Theory Workshop on Relativistic and QED Effects in Heavy Atoms	85-73790	0-88318-335-8
No. 137	Polymer-Flow Interaction (La Jolla Institute, 1985)	85-73915	0-88318-336-6
No. 138	Frontiers in Electronic Materials and Processing (Houston, TX, 1985)	86-70108	0-88318-337-4
No. 139	High-Current, High-Brightness, and High-Duty Factor Ion Injectors (La Jolla Institute, 1985)	86-70245	0-88318-338-2

No. 140	Boron-Rich Solids (Albuquerque, NM, 1985)	86-70246	0-88318-339-0
No. 141	Gamma-Ray Bursts (Stanford, CA, 1984)	86-70761	0-88318-340-4
No. 142	Nuclear Structure at High Spin, Excitation, and Momentum Transfer (Indiana University, 1985)	86-70837	0-88318-341-2
No. 143	Mexican School of Particles and Fields (Oaxtepec, México, 1984)	86-81187	0-88318-342-0
No. 144	Magnetospheric Phenomena in Astrophysics (Los Alamos, 1984)	86-71149	0-88318-343-9
No. 145	Polarized Beams at SSC & Polarized Antiprotons (Ann Arbor, MI & Bodega Bay, CA, 1985)	86-71343	0-88318-344-7
No. 146	Advances in Laser Science–I (Dallas, TX, 1985)	86-71536	0-88318-345-5
No. 147	Short Wavelength Coherent Radiation: Generation and Applications (Monterey, CA, 1986)	86-71674	0-88318-346-3
No. 148	Space Colonization: Technology and The Liberal Arts (Geneva, NY, 1985)	86-71675	0-88318-347-1
No. 149	Physics and Chemistry of Protective Coatings (Universal City, CA, 1985)	86-72019	0-88318-348-X
No. 150	Intersections Between Particle and Nuclear Physics (Lake Louise, Canada, 1986)	86-72018	0-88318-349-8
No. 151	Neural Networks for Computing (Snowbird, UT, 1986)	86-72481	0-88318-351-X
No. 152	Heavy Ion Inertial Fusion (Washington, DC, 1986)	86-73185	0-88318-352-8
No. 153	Physics of Particle Accelerators (SLAC Summer School, 1985) (Fermilab Summer School, 1984)	87-70103	0-88318-353-6
No. 154	Physics and Chemistry of Porous Media—II (Ridge Field, CT, 1986)	83-73640	0-88318-354-4